THE
TWENTY YEARS' CRISIS

1919 — 1939

22578

ʜᴀʀᴘᴇʀ 🔥 ᴛᴏʀᴄʜʙᴏᴏᴋs

*A reference-list of Harper Torchbooks, classified
by subjects, is printed at the end of this volume.*

THE
TWENTY YEARS' CRISIS
1919 — 1939

An Introduction to the Study of
International Relations

BY

EDWARD HALLETT CARR

HARPER TORCHBOOKS ♥ The Academy Library
Harper & Row, Publishers   New York

THE TWENTY YEARS' CRISIS
1919—1939

First HARPER TORCHBOOK edition published 1964 by
Harper & Row, Publishers, Incorporated
49 East 33rd Street
New York 16, N. Y.

TO THE MAKERS
OF THE COMING PEACE

" Philosophers make imaginary laws for imaginary commonwealths, and their discourses are as the stars which give little light because they are so high."

BACON, *On the Advancement of Learning*

" The roads to human power and to human knowledge lie close together and are nearly the same ; nevertheless, on account of the pernicious and inveterate habit of dwelling on abstractions, it is safer to begin and raise the sciences from those foundations which have relation to practice, and let the active part be as the seal which prints and determines the contemplative counterpart."

Id., Novum Organum

PREFACE TO SECOND EDITION

THE demand for a second edition of *The Twenty Years' Crisis* faced the author with a difficult decision. A work on international politics completed in the summer of 1939, however rigorously it eschewed prophecy, necessarily bore marks of its time in substance, in phraseology, in its use of tenses and, above all, in such phrases as " the War ", " post-War " and so forth, which can no longer be related without a strong effort on the part of the reader to the war of 1914–18. When, however, I approached the task of revision, it soon became clear that if I sought to re-write every passage which had been in some way affected by the subsequent march of events, I should be producing not a second edition of an old book but essentially a new one ; and this would have been a clumsy and unprofitable attempt to force new wine into old bottles. *The Twenty Years' Crisis* remains a study of the period between the two wars written as that period was coming to an end and must be treated on its merits as such. What I have done, therefore, is to recast phrases which would be misleading or difficult to readers now far remote in time from the original context, to modify a few sentences which have invited misunderstanding, and to remove two or three passages relating to current controversies which have been eclipsed or put in a different perspective by the lapse of time.

On the other hand, I have changed nothing of substance, and have not sought to modify expressions of opinion merely on the ground that I should not unreservedly endorse them to-day. Perhaps, therefore, I may be permitted to indicate here the two main respects in which I am conscious of having since departed to some degree from the outlook reflected in these pages.

In the first place, *The Twenty Years' Crisis* was written with the deliberate aim of counteracting the glaring and dangerous defect of nearly all thinking, both academic and popular, about international politics in English-speaking countries from 1919 to 1939 — the almost total neglect of the factor of power. To-day this defect, though it sometimes recurs when items of a future settlement are under discussion, has been to a con-

Preface to Second Edition

siderable extent overcome ; and some passages of *The Twenty Years' Crisis* state their argument with a rather one-sided emphasis which no longer seems as necessary or appropriate to-day as it did in 1939.

Secondly, the main body of the book too readily and too complacently accepts the existing nation-state, large or small, as the unit of international society, though the final chapter offers some reflexions, to which subsequent events have added point, on the size of the political and economic units of the future. The conclusion now seems to impose itself on any unbiassed observer that the small independent nation-state is obsolete or obsolescent and that no workable international organisation can be built on a membership of a multiplicity of nation-states. My present views on this point have been worked out in a small book recently published under the title *National-ism and After* ; and I can therefore with the better conscience take the only practicable course and leave the present work substantially as it was completed in 1939.

E. H. CARR

November 15, 1945

PREFACE TO FIRST EDITION

THIS book, which was originally planned in 1937, was sent to the press in the middle of July 1939 and had reached page proof when war broke out on September 3, 1939. To introduce into the text a few verbal modifications hastily made in the light of that event would have served little purpose; and I have accordingly preferred to leave it exactly as it was written at a time when war was already casting its shadow on the world, but when all hope of averting it was not yet lost. Wherever, therefore, such phrases as "the War", "pre-War" or "post-War" occur in the following pages, the reader will understand that the reference is to the War of 1914-18.

When the passions of war are aroused, it becomes almost fatally easy to attribute the catastrophe solely to the ambitions and the arrogance of a small group of men, and to seek no further explanation. Yet even while war is raging, there may be some practical importance in an attempt to analyse the underlying and significant, rather than the immediate and personal, causes of the disaster. If and when peace returns to the world, the lessons of the breakdown which has involved Europe in a second major war within twenty years and two months of the Versailles Treaty will need to be earnestly pondered. A settlement which, having destroyed the National Socialist rulers of Germany, leaves untouched the conditions which made the phenomenon of National Socialism possible, will run the risk of being as short-lived and as tragic as the settlement of 1919. No period of history will better repay study by the peacemakers of the future than the Twenty Years' Crisis which fills the interval between the two Great Wars. The next Peace Conference, if it is not to repeat the fiasco of the last, will have to concern itself with issues more fundamental than the drawing of frontiers. In this belief, I have ventured to dedicate this book to the makers of the coming peace.

The published sources from which I have derived help and inspiration are legion. I am specially indebted to two books which, though not specifically concerned with international relations, seem to me to have illuminated some of the fundamental problems of politics: Dr. Karl Mannheim's *Ideology and Utopia* and Dr. Reinhold Niebuhr's *Moral Man and Immoral Society.* Mr. Peter Drucker's *The End of Economic Man*, which did not come into my hands until my manuscript was virtually complete, contains some brilliant guesses and a most stimulating and suggestive diagnosis of the present crisis in world history. Many excellent historical and descriptive works about various aspects of international relations have appeared in the last twenty years, and my indebtedness to some of these is recorded in footnotes, which must take the place of a bibliography. But not one of these works known to me has attempted to analyse the profounder causes of the contemporary international crisis.

My obligations to individuals are still more extensive. In particular, I desire to record my deep gratitude to three friends who found time to read the whole of my manuscript, whose comments were equally stimulating whether they agreed or disagreed with my views, and whose suggestions are responsible for a great part of such value as this book possesses : Charles Manning, Professor of International Relations in the London School of Economics and Political Science ; Dennis Routh, Fellow of All Souls College, Oxford, and recently Lecturer in International Politics in the University College of Wales, Aberystwyth ; and a third, whose official position deprives me of the pleasure of naming him here. During the past three years I have been a member of a Study Group of the Royal Institute of International Affairs engaged on an enquiry into the problem of nationalism, the results of which are about to be published.[1] The lines of investigation pursued by this Group have sometimes touched or crossed those which I have been following in these pages ; and my colleagues in this Group and other contributors to its work have, in the course of our long discussions, unwittingly made numerous valuable contributions to the present book. To these, and to the many others who, in one way or another, consciously or unconsciously, have given me assistance and encouragement in the preparation of this volume, I tender my sincere thanks.

E. H. CARR

September 30, 1939

[1] *Nationalism :* A Study by a Group of Members of the Royal Institute of International Affairs (Oxford University Press).

CONTENTS

PART ONE

THE SCIENCE OF INTERNATIONAL POLITICS

PART TWO

THE INTERNATIONAL CRISIS

PART THREE

POLITICS, POWER AND MORALITY

Contents

PART FOUR
LAW AND CHANGE

CONCLUSION

THE
TWENTY YEARS' CRISIS

1919 — 1939

THE SCIENCE OF INTERNATIONAL POLITICS

CHAPTER I

THE BEGINNINGS OF A SCIENCE

THE science of international politics is in its infancy. Down to 1914, the conduct of international relations was the concern of persons professionally engaged in it. In democratic countries, foreign policy was traditionally regarded as outside the scope of party politics ; and the representative organs did not feel themselves competent to exercise any close control over the mysterious operations of foreign offices. In Great Britain, public opinion was readily aroused if war occurred in any region traditionally regarded as a sphere of British interest, or if the British navy momentarily ceased to possess that margin of superiority over potential rivals which was then deemed essential. In continental Europe, conscription and the chronic fear of foreign invasion had created a more general and continuous popular awareness of international problems. But this awareness found expression mainly in the labour movement, which from time to time passed somewhat academic resolutions against war. The constitution of the United States of America contained the unique provision that treaties were concluded by the President " by and with the advice and consent of the Senate ". But the foreign relations of the United States seemed too parochial to lend any wider significance to this exception. The more picturesque aspects of diplomacy had a certain news value. But nowhere, whether in universities or in wider intellectual circles, was there organised study of current international affairs. War was still regarded mainly as the business of soldiers ; and the corollary of this was that international politics were the business of diplomats. There was no general desire to take the conduct of international affairs out of the hands of the professionals or even to pay serious and systematic attention to what they were doing.

The Science of International Politics

The war of 1914–18 made an end of the view that war is a matter which affects only professional soldiers and, in so doing, dissipated the corresponding impression that international politics could safely be left in the hands of professional diplomats. The campaign for the popularisation of international politics began in the English-speaking countries in the form of an agitation against secret treaties, which were attacked, on insufficient evidence, as one of the causes of the war. The blame for the secret treaties should have been imputed, not to the wickedness of the governments, but to the indifference of the peoples. Everybody knew that such treaties were concluded. But before the war of 1914 few people felt any curiosity about them or thought them objectionable.[1] The agitation against them was, however, a fact of immense importance. It was the first symptom of the demand for the popularisation of international politics and heralded the birth of a new science.

Purpose and Analysis in Political Science

The science of international politics has, then, come into being in response to a popular demand. It has been created to serve a purpose and has, in this respect, followed the pattern of other sciences. At first sight, this pattern may appear illogical. Our first business, it will be said, is to collect, classify and analyse our facts and draw our inferences ; and we shall then be ready to investigate the purpose to which our facts and our deductions can be put. The processes of the human mind do not, however, appear to develop in this logical order. The human mind works, so to speak, backwards. Purpose, which should logically follow analysis, is required to give it both its initial impulse and its direction. " If society has a technical need ", wrote Engels, " it serves as a greater spur to the progress of science than do ten universities." [2] The first extant text-book of geometry " lays down an aggregate of

---

[1] A recent historian of the Franco-Russian alliance, having recorded the protests of a few French radicals against the secrecy which enveloped this transaction, continues : " Parliament and opinion tolerated this complete silence, and were content to remain in absolute ignorance of the provisions and scope of the agreement " (Michon, *L'Alliance Franco-Russe*, p. 75). In 1898, in the Chamber of Deputies, Hanotaux was applauded for describing the disclosure of its terms as " absolutely impossible " (*ibid*. p. 82).

[2] Quoted in Sidney Hook, *Towards the Understanding of Karl Marx*, p. 279.

practical rules designed to solve concrete problems : ' rule for measuring a round fruitery ' ; ' rule for laying out a field ' ; ' computation of the fodder consumed by geese and oxen ' ".[1] Reason, says Kant, must approach nature " not . . . in the character of a pupil, who listens to all that his master chooses to tell him, but in that of a judge, who compels the witnesses to reply to those questions which he himself thinks fit to propose ".[2] " We cannot study even stars or rocks or atoms ", writes a modern sociologist, " without being somehow determined, in our modes of systematisation, in the prominence given to one or another part of our subject, in the form of the questions we ask and attempt to answer, by direct and human interests." [3] It is the purpose of promoting health which creates medical science, and the purpose of building bridges which creates the science of engineering. Desire to cure the sicknesses of the body politic has given its impulse and its inspiration to political science. Purpose, whether we are conscious of it or not, is a condition of thought ; and thinking for thinking's sake is as abnormal and barren as the miser's accumulation of money for its own sake. " The wish is father to the thought " is a perfectly exact description of the origin of normal human thinking.

If this is true of the physical sciences, it is true of political science in a far more intimate sense. In the physical sciences, the distinction between the investigation of facts and the purpose to which the facts are to be put is not only theoretically valid, but is constantly observed in practice. The laboratory worker engaged in investigating the causes of cancer may have been originally inspired by the purpose of eradicating the disease. But this purpose is in the strictest sense irrelevant to the investigation and separable from it. His conclusion can be nothing more than a true report on facts. It cannot help to make the facts other than they are ; for the facts exist independently of what anyone thinks about them. In the political sciences, which are concerned with human behaviour, there are no such facts. The investigator is inspired by the desire to cure some ill of the body politic. Among the causes of the trouble, he diagnoses the fact that human beings normally react to

[1] J. Rueff, *From the Physical to the Social Sciences* (Engl. transl.), p. 27.
[2] Kant, *Critique of Pure Reason* (Everyman ed.), p. 11.
[3] MacIver, *Community*, p. 56.

certain conditions in a certain way. But this is not a fact comparable with the fact that human bodies react in a certain way to certain drugs. It is a fact which may be changed by the desire to change it ; and this desire, already present in the mind of the investigator, may be extended, as the result of his investigation, to a sufficient number of other human beings to make it effective. The purpose is not, as in the physical sciences, irrelevant to the investigation and separable from it : it is itself one of the facts. In theory, the distinction may no doubt still be drawn between the role of the investigator who establishes the facts and the role of the practitioner who considers the right course of action. In practice, one role shades imperceptibly into the other. Purpose and analysis become part and parcel of a single process.

A few examples will illustrate this point. Marx, when he wrote *Capital*, was inspired by the purpose of destroying the capitalist system just as the investigator of the causes of cancer is inspired by the purpose of eradicating cancer. But the facts about capitalism are not, like the facts about cancer, independent of the attitude of people towards it. Marx's analysis was intended to alter, and did in fact alter, that attitude. In the process of analysing the facts, Marx altered them To attempt to distinguish between Marx the scientist and Marx the propagandist is idle hair-splitting. The financial experts, who in the summer of 1932 advised the British Government that it was possible to convert 5 per cent War Loan at the rate of $3\frac{1}{2}$ per cent, no doubt based their advice on an analysis of certain facts ; but the fact that they gave this advice was one of the facts which, being known to the financial world, made the operation successful. Analysis and purpose were inextricably blended. Nor is it only the thinking of professional or qualified students of politics which constitutes a political fact. Everyone who reads the political columns of a newspaper or attends a political meeting or discusses politics with his neighbour is to that extent a student of politics ; and the judgment which he forms becomes (especially, but not exclusively, in democratic countries) a factor in the course of political events. Thus a reviewer might conceivably criticise this book on the ground, not that it was false, but that it was inopportune ; and this criticism, whether justified or not, would be intelligible, whereas the same criticism of a book about the causes of cancer would be meaningless. Every political judgment helps to modify the facts on which it is passed. Political thought is itself a form of political action. Political science is the science not only of what is, but of what ought to be.

The Role of Utopianism

If therefore purpose precedes and conditions thought, it is not surprising to find that, when the human mind begins to exercise itself in some fresh field, an initial stage occurs in which the element of wish or purpose is overwhelmingly strong, and the inclination to analyse facts and means weak or non-existent. Hobhouse notes as a characteristic of " the most primitive peoples " that " the evidence of the truth of an idea is not yet separate from the quality which renders it pleasant ".[1] The same would appear to be conspicuously true of the primitive, or " utopian ", stage of the political sciences. During this stage, the investigators will pay little attention to existing " facts " or to the analysis of cause and effect, but will devote themselves whole-heartedly to the elaboration of visionary projects for the attainment of the ends which they have in view— projects whose simplicity and perfection give them an easy and universal appeal. It is only when these projects break down, and wish or purpose is shewn to be incapable by itself of achieving the desired end, that the investigators will reluctantly call in the aid of analysis, and the study, emerging from its infantile and utopian period, will establish its claim to be regarded as a science. " Sociology ", remarks Professor Ginsberg, " may be said to have arisen by way of reaction against sweeping generalisations unsupported by detailed inductive enquiry." [2]

It may not be fanciful to find an illustration of this rule even in the domain of physical science. During the Middle Ages, gold was a recognised medium of exchange. But economic relations were not sufficiently developed to require more than a limited amount of such a medium. When the new economic conditions of the fourteenth and fifteenth centuries introduced a widespread system of money transactions, and the supply of gold was found to be inadequate for the purpose, the

[1] L. T. Hobhouse, *Development and Purpose*, p. 100.
[2] M. Ginsberg, *Sociology*, p. 25.

wise men of the day began to experiment in the possibility of transmuting commoner metals into gold. The thought of the alchemist was purely purposive. He did not stop to enquire whether the properties of lead were such as to make it transmutable into gold. He assumed that the end was absolute (i.e. that gold must be produced), and that means and material must somehow be adapted to it. It was only when this visionary project ended in failure that the investigators were prompted to apply their thought to an examination of " facts ", i.e. the nature of matter ; and though the initial utopian purpose of making gold out of lead is probably as far as ever from fulfilment, modern physical science has been evolved out of this primitive aspiration.

Other illustrations may be taken from fields more closely akin to our present subject.

It was in the fifth and fourth centuries B.C. that the first serious recorded attempts were made to create a science of politics. These attempts were made independently in China and in Greece. But neither Confucius nor Plato, though they were of course profoundly influenced by the political institutions under which they lived, really tried to analyse the nature of those institutions or to seek the underlying causes of the evils which they deplored. Like the alchemists, they were content to advocate highly imaginative solutions whose relation to existing facts was one of flat negation.[1] The new political order which they propounded was as different from anything they saw around them as gold from lead. It was the product not of analysis, but of aspiration.

In the eighteenth century, trade in Western Europe had become so important as to render irksome the innumerable restrictions placed on it by governmental authority and justified by mercantilist theory. The protest against these restrictions took the form of a wishful vision of universal free trade ; and out of this vision the physiocrats in France, and Adam Smith in Great Britain, created a science of political economy. The new science was based primarily on a negation of existing reality and on certain artificial and unverified generalisations

[1] " Plato and Plotinus, More and Campanella constructed their fanciful societies with those materials which were omitted from the fabric of the actual communities by the defects of which they were inspired. The Republic, the Utopia, and the City of the Sun were protests against a state of things which the experience of their authors taught them to condemn " (Acton, *History of Freedom*, p. 270).

about the behaviour of a hypothetical economic man. In practice, it achieved some highly useful and important results. But economic theory long retained its utopian character ; and even to-day some " classical economists " insist on regarding universal free trade — an imaginary condition which has never existed — as the normal postulate of economic science, and all reality as a deviation from this utopian prototype.[1]

In the opening years of the nineteenth century, the industrial revolution created a new social problem to engage human thought in Western Europe. The pioneers who first set out to tackle this problem were the men on whom posterity has bestowed the name of " utopian socialists " : Saint-Simon and Fourier in France, Robert Owen in England. These men did not attempt to analyse the nature of class-interests or class-consciousness or of the class-conflict to which they gave rise. They simply made unverified assumptions about human behaviour and, on the strength of these, drew up visionary schemes of ideal communities in which men of all classes would live together in amity, sharing the fruits of their labours in proportion to their needs. For all of them, as Engels remarked, " socialism is the expression of absolute truth, reason and justice, and needs only be discovered in order to conquer all the world in virtue of its own power ".[2] The utopian socialists did valuable work in making men conscious of the problem and of the need of tackling it. But the solution propounded by them had no logical connexion with the conditions which created the problem. Once more, it was the product not of analysis, but of aspiration.

Schemes elaborated in this spirit would not, of course, work.

[1] " L'économie politique libérale a été un des meilleurs exemples d'utopies qu'on puisse citer. On avait imaginé une société où tout serait ramené à des types commerciaux, sous la loi de la plus complète concurrence ; on reconnaît aujourd'hui que cette société idéale serait aussi difficile à realiser que celle de Platon " (Sorel, *Réflexions sur la violence*, p. 47). Compare Professor Robbins' well-known defence of *laissez-faire* economics : " The idea of a co-ordination of human activity by means of a system of impersonal rules, within which what spontaneous relations arise are conducive to mutual benefit, is a conception at least as subtle, at least as ambitious, as the conception of prescribing each action or each type of action by a central planning authority ; and it is perhaps not less in harmony with the requirements of a spiritually sound society " (*Economic Planning and International Order*, p. 229). It would be equally true, and perhaps equally useful, to say that the constitution of Plato's Republic is at least as subtle, ambitious and satisfying to spiritual requirements as that of any state which has ever existed.

[2] Engels, *Socialism, Utopian and Scientific* (Engl. transl.), p. 26.

Just as nobody has ever been able to make gold in a laboratory, so nobody has ever been able to live in Plato's republic or in a world of universal free trade or in Fourier's phalansteries. But it is, nevertheless, perfectly right to venerate Confucius and Plato as the founders of political science, Adam Smith as the founder of political economy, and Fourier and Owen as the founders of socialism. The initial stage of aspiration towards an end is an essential foundation of human thinking. The wish is father to the thought. Teleology precedes analysis.

The teleological aspect of the science of international politics has been conspicuous from the outset. It took its rise from a great and disastrous war ; and the overwhelming purpose which dominated and inspired the pioneers of the new science was to obviate a recurrence of this disease of the international body politic. The passionate desire to prevent war determined the whole initial course and direction of the study. Like other infant sciences, the science of international politics has been markedly and frankly utopian. It has been in the initial stage in which wishing prevails over thinking, generalisation over observation, and in which little attempt is made at a critical analysis of existing facts or available means. In this stage, attention is concentrated almost exclusively on the end to be achieved. The end has seemed so important that analytical criticism of the means proposed has too often been branded as destructive and unhelpful. When President Wilson, on his way to the Peace Conference, was asked by some of his advisers whether he thought his plan of a League of Nations would work, he replied briefly : " If it won't work, it must be made to work ".[1] The advocate of a scheme for an international police force or for " collective security ", or of some other project for an international order, generally replied to the critic not by an argument designed to shew how and why he thought his plan will work, but either by a statement that it must be made to work because the consequences of its failure to work would be so disastrous, or by a demand for some alternative nostrum.[2] This must be the spirit in which the alchemist or

---

[1] R. S. Baker, *Woodrow Wilson and World Settlement*, i. p. 93.

[2] " There is the old well-known story about the man who, during the Lisbon earthquake of 1775, went about hawking anti-earthquake pills ; but one incident is forgotten — when someone pointed out that the pills could not possibly be of use, the hawker replied : ' But what would you put in their place ? ' " (L. B. Namier, *In the Margin of History*, p. 20).

the utopian socialist would have answered the sceptic who questioned whether lead could be turned into gold or men made to live in model communities. Thought has been at a discount. Much that was said and written about international politics between 1919 and 1939 merited the stricture applied in another context by the economist Marshall, who compares " the nervous irresponsibility which conceives hasty utopian schemes " to the " bold facility of the weak player who will speedily solve the most difficult chess problem by taking on himself to move the black men as well as the white ".[1] In extenuation of this intellectual failure, it may be said that, during the earlier of these years, the black pieces in international politics were in the hands of such weak players that the real difficulties of the game were scarcely manifest even to the keenest intelligence. The course of events after 1931 clearly revealed the inadequacy of pure aspiration as the basis for a science of international politics, and made it possible for the first time to embark on serious critical and analytical thought about international problems.

The Impact of Realism

No science deserves the name until it has acquired sufficient humility not to consider itself omnipotent, and to distinguish the analysis of what is from aspiration about what should be. Because in the political sciences this distinction can never be absolute, some people prefer to withhold from them the right to the title of science. In both physical and political sciences, the point is soon reached where the initial stage of wishing must be succeeded by a stage of hard and ruthless analysis. The difference is that political sciences can never wholly emancipate themselves from utopianism, and that the political scientist is apt to linger for a longer initial period than the physical scientist in the utopian stage of development. This is perfectly natural. For while the transmutation of lead into gold would be no nearer if everyone in the world passionately desired it, it is undeniable that if everyone really desired a " world-state " or " collective security " (and meant the same thing by those terms), it would be easily attained ; and the student of international politics may be forgiven if he begins by supposing that his task is to make everyone desire it. It takes him some

[1] *Economic Journal* (1907), xvii. p. 9.

time to understand that no progress is likely to be made along this path, and that no political utopia will achieve even the most limited success unless it grows out of political reality. Having made the discovery, he will embark on that hard ruthless analysis of reality which is the hallmark of science ; and one of the facts whose causes he will have to analyse is the fact that few people do desire a " world-state " or " collective security ", and that those who think they desire it mean different and incompatible things by it. He will have reached a stage when purpose by itself is seen to be barren, and when analysis of reality has forced itself upon him as an essential ingredient of his study.

The impact of thinking upon wishing which, in the development of a science, follows the breakdown of its first visionary projects, and marks the end of its specifically utopian period, is commonly called realism. Representing a reaction against the wish-dreams of the initial stage, realism is liable to assume a critical and somewhat cynical aspect. In the field of thought, it places its emphasis on the acceptance of facts and on the analysis of their causes and consequences. It tends to depreciate the role of purpose and to maintain, explicitly or implicitly, that the function of thinking is to study a sequence of events which it is powerless to influence or to alter. In the field of action, realism tends to emphasise the irresistible strength of existing forces and the inevitable character of existing tendencies, and to insist that the highest wisdom lies in accepting, and adapting oneself to, these forces and these tendencies. Such an attitude, though advocated in the name of " objective " thought, may no doubt be carried to a point where it results in the sterilisation of thought and the negation of action. But there is a stage where realism is the necessary corrective to the exuberance of utopianism, just as in other periods utopianism must be invoked to counteract the barrenness of realism. Immature thought is predominantly purposive and utopian. Thought which rejects purpose altogether is the thought of old age. Mature thought combines purpose with observation and analysis. Utopia and reality are thus the two facets of political science. Sound political thought and sound political life will be found only where both have their place.

CHAPTER 2

UTOPIA AND REALITY

THE antithesis of utopia and reality — a balance always swinging towards and away from equilibrium and never completely attaining it — is a fundamental antithesis revealing itself in many forms of thought. The two methods of approach — the inclination to ignore what was and what is in contemplation of what should be, and the inclination to deduce what should be from what was and what is — determine opposite attitudes towards every political problem. " It is the eternal dispute ", as Albert Sorel puts it, " between those who imagine the world to suit their policy, and those who arrange their policy to suit the realities of the world." [1] It may be suggestive to elaborate this antithesis before proceeding to an examination of the current crisis of international politics.

Free Will and Determination

The antithesis of utopia and reality can in some aspects be identified with the antithesis of Free Will and Determinism. The utopian is necessarily voluntarist : he believes in the possibility of more or less radically rejecting reality, and substituting his utopia for it by an act of will. The realist analyses a predetermined course of development which he is powerless to change. For the realist, philosophy, in the famous words of Hegel's preface to his *Philosophy of Right*, always " comes too late " to change the world. By means of philosophy, the old order " cannot be rejuvenated, but only known ". The utopian, fixing his eyes on the future, thinks in terms of creative spontaneity : the realist, rooted in the past, in terms of causality. All healthy human action, and therefore all healthy thought, must establish a balance between utopia and reality, between free will and determinism. The complete realist, unconditionally accepting the causal sequence of events, deprives himself of the possibility of changing reality. The complete utopian, by rejecting the causal sequence, deprives himself of the possibility

[1] A. Sorel, *L'Europe et la Révolution Française*, p. 474.

of understanding either the reality which he is seeking to change or the processes by which it can be changed. The characteristic vice of the utopian is naivety ; of the realist, sterility.[1]

Theory and Practice

The antithesis of utopia and reality also coincides with the antithesis of theory and practice. The utopian makes political theory a norm to which political practice ought to conform. The realist regards political theory as a sort of codification of political practice. The relationship of theory and practice has come to be recognised in recent years as one of the central problems of political thought. Both the utopian and the realist distort this relationship. The utopian, purporting to recognise the interdependence of purpose and fact, treats purpose as if it were the only relevant fact, and constantly couches optative propositions in the indicative mood. The American Declaration of Independence maintains that " all men are created equal ", Mr. Litvinov that " peace is indivisible ",[2] and Sir Norman Angell that " the biological division of mankind into independent warring states " is a " scientific ineptitude ".[3] Yet it is a matter of common observation that all men are not born equal even in the United States, and that the Soviet Union can remain at peace while its neighbours are at war ; and we should probably think little of a zoologist who described a man-eating tiger as a " scientific ineptitude ". These propositions are items in a political programme disguised as statements of fact ;[4] and the utopian inhabits a dream-world of such " facts ", remote from the world of reality where quite contrary facts may be observed. The realist has no difficulty in perceiving that these utopian propositions are not facts but aspirations, and belong to the optative not to the indicative mood ;

[1] The psychologist may be interested to trace here an analogy — it would be dangerous to treat it as more — with Jung's classification of psychological types as "introverted" and "extraverted" (Jung, *Psychological Types*) or William James's pairs of opposites : Rationalist-Empiricist, Intellectualist-Sensationalist, Idealist-Materialist, Optimistic-Pessimistic, Religious-Irreligious, Free-willist-Fatalistic, Monistic-Pluralistic, Dogmatical-Sceptical (W. James, Pragmatism).

[2] *League of Nations : Sixteenth Assembly*, p. 72.

[3] Angell, *The Great Illusion*, p. 138.

[4] Similarly, Marx's theory of surplus value has, in the words of a sympathetic critic, " rather the significance of a political and social slogan than of an economic truth " (M. Beer, *The Life and Teaching of Karl Marx*, p. 129).

and he goes on to shew that, considered as aspirations, they are not *a priori* propositions, but are rooted in the world of reality in a way which the utopian altogether fails to understand. Thus for the realist, the equality of man is the ideology of the under-privileged seeking to raise themselves to the level of the privileged ; the indivisibility of peace the ideology of states which, being particularly exposed to attack, are eager to establish the principle that an attack on them is a matter of concern to other states more fortunately situated ; [1] the ineptitude of sovereign states the ideology of predominant Powers which find the sovereignty of other states a barrier to the enjoyment of their own predominant position. This exposure of the hidden foundations of utopian theory is a necessary preliminary to any serious political science. But the realist, in denying any *a priori* quality to political theories, and in proving them to be rooted in practice, falls easily into a determinism which argues that theory, being nothing more than a rationalisation of conditioned and predetermined purpose, is a pure excrescence and impotent to alter the course of events. While therefore the utopian treats purpose as the sole ultimate fact, the realist runs the risk of treating purpose merely as the mechanical product of other facts. If we recognise that this mechanisation of human will and human aspiration is untenable and intolerable, then we must recognise that theory, as it develops out of practice and develops into practice, plays its own transforming role in the process. The political process does not consist, as the realist believes, purely in a succession of phenomena governed by mechanical laws of causation ; nor does it consist, as the utopian believes, purely in the application to practice of certain theoretical truths evolved out of their inner consciousness by wise and far-seeing people. Political science must be based on a recognition of the interdependence of theory and practice, which can be attained only through a combination of utopia and reality.

The Intellectual and the Bureaucrat

A concrete expression of the antithesis of theory and practice in politics is the opposition between the " intellectual " and the

[1] Having discovered that other states were perhaps more open to attack than themselves, the Soviet authorities in May 1939 dismissed Mr. Litvinov and ceased to talk about the indivisibility of peace.

" bureaucrat ",[1] the former trained to think mainly on *a priori* lines, the latter empirically. It is in the nature of things that the intellectual should find himself in the camp which seeks to make practice conform to theory ; for intellectuals are particularly reluctant to recognise their thought as conditioned by forces external to themselves, and like to think of themselves as leaders whose theories provide the motive force for so-called men of action. Moreover, the whole intellectual outlook of the last two hundred years has been strongly coloured by the mathematical and natural sciences. To establish a general principle, and to test the particular in the light of that principle, has been assumed by most intellectuals to be the necessary foundation and starting-point of any science. In this respect, utopianism with its insistence on general principles may be said to represent the characteristic intellectual approach to politics. Woodrow Wilson, the most perfect modern example of the intellectual in politics, " excelled in the exposition of fundamentals. . . His political method . . . was to base his appeal upon broad and simple principles, avoiding commitment upon specific measures." [2] Some supposedly general principle, such as " national self-determination ", " free trade " or " collective security " (all of which will be easily recognised by the realist as concrete expressions of particular conditions and interests), is taken as an absolute standard, and policies are adjudged good or bad by the extent to which they conform to, or diverge from, it. In modern times, intellectuals have been the leaders of every utopian movement ; and the services which utopianism has rendered to political progress must be credited in large part to them. But the characteristic weakness of utopianism is also the characteristic weakness of the political intellectuals — failure to understand existing reality and the way in which their own standards are rooted in it. " They could give to their political aspirations ", wrote Meinecke of the role of intellectuals in German politics, " a spirit of purity and independence, of philosophical idealism and of elevation

---

[1] The term " bureaucrat " may be taken for this purpose to include those members of the fighting services who are concerned with the direction of policy. It is, perhaps, unnecessary to add that not every possessor of an intellect is an intellectual, or every occupant of a desk in a government department a bureaucrat. There are, nevertheless, modes of thought which are, broadly speaking, characteristic of the " bureaucrat " and the " intellectual " respectively.

[2] R. S. Baker, *Woodrow Wilson : Life and Letters*, iii. p. 90.

above the concrete play of interests . . . but through their defective feeling for the realistic interests of actual state life they quickly descended from the sublime to the extravagant and eccentric." [1]

It has often been argued that the intellectuals are less directly conditioned in their thinking than those groups whose coherence depends on a common economic interest, and that they therefore occupy a vantage-point *au-dessus de la mêlée*. As early as 1905, Lenin attacked " the old-fashioned view of the intelligentsia as capable . . . of standing outside class ".[2] More recently, this view has been resuscitated by Dr. Mannheim, who argues that the intelligentsia, being " *relatively* classless " and " socially unattached ", " subsumes in itself all those interests with which social life is permeated ", and can therefore attain a higher measure of impartiality and objectivity.[3] In a certain limited sense, this is true. But any advantage derived from it would seem to be nullified by a corresponding disability, i.e. detachment from the masses whose attitude is the determining factor in political life. Even where the illusion of their leadership was strongest, modern intellectuals have often found themselves in the position of officers whose troops were ready enough to follow them in quiet times, but could be relied on to desert in any serious engagement. In Germany and many smaller European countries, the democratic constitutions of 1919 were the work of devoted intellectuals, and achieved a high degree of theoretical perfection. But when a crisis occurred, they broke down almost everywhere through failure to win the durable allegiance of the mass of the population. In the United States, the intellectuals played a preponderant part in creating the League of Nations, and most of them remained avowed supporters of it. Yet the mass of the American people, having appeared to follow their lead, rejected it when the critical moment arrived. In Great Britain, the intellectuals secured, by a devoted and energetic propaganda, overwhelming paper support for the League of Nations. But when the Covenant appeared to require action which might have entailed practical consequences for the mass of the people, successive governments preferred inaction ; and the protests of the in-

[1] Meinecke, *Staat und Persönlichkeit*, p. 136.
[2] Lenin, *Works* (2nd Russian ed.), vii. p. 72.
[3] Mannheim, *Ideology and Utopia*, pp. 137-40.

tellectuals caused no perceptible reaction in the country.

The bureaucratic approach to politics is, on the other hand, fundamentally empirical. The bureaucrat purports to handle each particular problem " on its merits ", to eschew the formulation of principles and to be guided on the right course by some intuitive process born of long experience and not of conscious reasoning. " There are no general cases ", said a French official, acting as French Delegate at an Assembly of the League of Nations ; " there are only specific cases." [1] In his dislike of theory, the bureaucrat resembles the man of action. " On s'engage, puis on voit " is a motto attributed to more than one famous general. The excellence of the British civil service is partly due to the ease with which the bureaucratic mentality accommodates itself to the empirical tradition of British politics. The perfect civil servant conforms closely to the popular picture of the English politician as a man who recoils from written constitutions and solemn covenants, and lets himself be guided by precedent, by instinct, by feel for the right thing. This empiricism is itself, no doubt, conditioned by a specific point of view, and reflects the conservative habit of English political life. The bureaucrat, perhaps more explicitly than any other class of the community, is bound up with the existing order, the maintenance of tradition, and the acceptance of precedent as the " safe " criterion of action. Hence bureaucracy easily degenerates into the rigid and empty formalism of the mandarin, and claims an esoteric understanding of appropriate procedures which is not accessible even to the most intelligent outsider. " Expérience vaut mieux que science " is the typical bureaucratic motto. " Attainments in learning and science ", wrote Bryce, voicing a widely felt prejudice, " do little to make men wise in politics." [2] When a bureaucrat wishes to damn a proposal, he calls it " academic ". Practice, not theory, bureaucratic training, not intellectual brilliance, is the school of political wisdom. The bureaucrat tends to make politics an end in themselves. It is worth remarking that both Machiavelli and Bacon were bureaucrats.

This fundamental antithesis between intellectual and bureaucratic modes of thought, always and everywhere latent, has appeared in the last half century in a quarter where it would

---

[1] *League of Nations : Fifteenth Assembly*, Sixth Committee, p. 62.
[2] Bryce, *Modern Democracies*, i. p. 89.

hardly have been looked for : in the labour movement. Writing in the 1870's, Engels congratulated the German workers on the fact that they " belong to the most theoretical nation in the world, and have retained that theoretical sense which has been almost completely lost by the so-called ' educated ' classes in Germany ". He contrasted this happy state with " the indifference to all theory which is one of the chief reasons of the slow progress of the English workers' movement ".[1] Forty years later, another German writer confirmed this observation.[2] The theoretical analysis of Marxist doctrine became one of the principal preoccupations of leading German Social Democrats ; and many observers believe that this one-sided intellectual development was an important factor in the ultimate collapse of the party. The British labour movement, until the last few years, entirely eschewed theory. At present, imperfect harmony between the intellectual and trade union wings is a notorious source of embarrassment to the Labour Party. The trade unionist tends to regard the intellectual as a utopian theorist lacking experience in the practical problems of the movement. The intellectual condemns the trade union leader as a bureaucrat. The recurrent conflicts between factions within the Bolshevik party in Soviet Russia were in part, at any rate, explicable as conflicts between the " party intelligentsia ", represented by Bukharin, Kamenev, Radek and Trotsky, and the " party machine " represented by Lenin, Sverdlov (till his death in 1919) and Stalin.[3]

The opposition between intellectual and bureaucrat was particularly prominent in Great Britain during the twenty years between the wars in the field of foreign affairs. During the first world war, the Union of Democratic Control, an organisation of utopian intellectuals, strove to popularise the view that the war was largely due to the control of foreign affairs in all countries by professional diplomats. Woodrow

---

[1] Quoted in Lenin, *Works* (2nd Russian ed.), iv. p. 381.

[2] " We possess the most theoretical labour movement in the world " (F. Naumann, *Central Europe*, Engl. transl., p. 121).

[3] This interpretation, which appears in Mirsky's *Lenin* (pp. 111, 117-18) published in 1931, received further confirmation from subsequent events. The rift went back to the earliest days of the party. In 1904 Lenin was attacking the party intellectuals for their neglect of discipline and organisation, and the intellectuals were attacking Lenin's bureaucratic methods (Lenin, *Works* (2nd Russian ed.), vi. pp. 309-11).

Wilson believed that peace would be secured if international issues were settled " not by diplomats or politicians each eager to serve his own interests, but by dispassionate scientists — geographers, ethnologists, economists — who had made studies of the problems involved ".[1] Bureaucrats, and especially diplomats, were long regarded with suspicion in League of Nations circles ; and it was considered that the League would contribute greatly to the solution of international problems by taking them out of the reactionary hands of foreign offices. Wilson, in introducing the draft Covenant to the plenary session of the Peace Conference, spoke of " the feeling that, if the deliberating body of the League of Nations was merely to be a body of officials representing the various governments, the peoples of the world would not be sure that some of the mistakes which preoccupied officials had admittedly made might not be repeated ".[2] Later, in the House of Commons, Lord Cecil was more scathing :

> I am afraid I came to the conclusion at the Peace Conference, from my own experience, that the Prussians were not exclusively confined to Germany. There is also the whole tendency and tradition of the official classes. . . . You cannot avoid the conclusion that there is a tendency among them to think that whatever is is right.[3]

At the Second Assembly, Lord Cecil invoked the support of " public opinion ", which the League was supposed to represent, against the " official classes " ; [4] and such appeals were frequently heard during the next ten years. The bureaucrat for his part equally mistrusted the missionary zeal of enthusiastic intellectuals for collective security, world order, and general disarmament — schemes which seemed to him the product of pure theory divorced from practical experience. The disarmament issue well illustrated this divergence of view. For the intellectual, the general principle was simple and straightforward ; the alleged difficulties of applying it were due to obstruction by the " experts ".[5] For the expert, the general

[1] R. S. Baker, *Woodrow Wilson and World Settlement*, i. p. 112.
[2] *History of the Peace Conference*, ed. H. Temperley, iii. p. 62.
[3] House of Commons, July 21, 1919 : *Official Report*, col. 993.
[4] *League of Nations : Second Assembly*, Third Committee, p. 281.
[5] " It is not to be tolerated ", said the Belgian socialist De Brouckère, " that the people should be robbed of their hopes of peace by experts who are losing themselves in the maze of technicalities which, with a little goodwill, might be

principle was meaningless and utopian ; whether armaments could be reduced, and if so which, was a " practical " question to be decided in each case "on its merits ".

Left and Right

The antithesis of utopia and reality, and of theory and practice, further reproduces itself in the antithesis of radical and conservative, of Left and Right, though it would be rash to assume that parties carrying these labels always represent these underlying tendencies. The radical is necessarily utopian, and the conservative realist. The intellectual, the man of theory, will gravitate towards the Left just as naturally as the bureaucrat, the man of practice, will gravitate towards the Right. Hence the Right is weak in theory, and suffers through its inaccessibility to ideas. The characteristic weakness of the Left is failure to translate its theory into practice — a failure for which it is apt to blame the bureaucrats, but which is inherent in its utopian character. " The Left has reason (*Vernunft*), the Right has wisdom (*Verstand*) ", wrote the Nazi philosopher, Moeller van den Bruck.[1] From the days of Burke onwards, English conservatives have always strongly denied the possibility of deducing political practice by a logical process from political theory. " To follow the syllogism alone is a short cut to the bottomless pit ", says Lord Baldwin [2] — a phrase which may suggest that he practises as well as preaches abstention from rigorously logical modes of thought. Mr. Churchill refuses to believe that "extravagant logic in doctrine " appeals to the British elector.[3] A particularly clear definition of different attitudes towards foreign policy comes from a speech made in the House of Commons by Neville Chamberlain in answer to a Labour critic :

What does the hon. Member mean by foreign policy ? You can lay down sound and general propositions. You

disentangled in a few hours " (Peace and Disarmament Committee of the Women's International Organisations : Circular of May 15, 1932). About the same time, Lord Cecil was reported in the same sense : " If the matter was to be left to experts nothing would be done. They were, he was sure, most able, conscientious, highly instructed gentlemen, but just look at their training " (*Manchester Guardian*, May 18, 1932).

[1] Moeller van den Bruck, *Das Dritte Reich* (3rd ed.), p. 257.
[2] Baldwin, *On England*, p. 153.   [3] Winston Churchill, *Step by Step*, p. 147.

can say that your foreign policy is to maintain peace ; you can say that it is to protect British interests, you can say that it is to use your influence, such as it is, on behalf of the right against the wrong, as far as you can tell the right from the wrong. You can lay down all these general principles, but that is not a policy. Surely, if you are to have a policy you must take the particular situations and consider what action or inaction is suitable for those particular situations. That is what I myself mean by policy, and it is quite clear that as the situations and conditions in foreign affairs continually change from day to day, your policy cannot be stated once and for all, if it is to be applicable to every situation that arises.[1]

The intellectual superiority of the Left is seldom in doubt The Left alone thinks out principles of political action and evolves ideals for statesmen to aim at. But it lacks practical experience which comes from close contact with reality. In Great Britain after 1919, it was a serious misfortune that the Left, having enjoyed office for negligible periods, had little experience of administrative realities and became more and more a party of pure theory, while the Right, having spent so little time in opposition, had few temptations to pit the perfection of theory against the imperfections of practice. In Soviet Russia, the group in power is more and more discarding theory in favour of practice as it loses the memory of its revolutionary origin. History everywhere shews that, when Left parties or politicians are brought into contact with reality through the assumption of political office, they tend to abandon their " doctrinaire " utopianism and move towards the Right, often retaining their Left labels and thereby adding to the confusion of political terminology.

Ethics and Politics

Most fundamental of all, the antithesis of utopia and reality is rooted in a different conception of the relationship of politics and ethics. The antithesis between the world of value and the world of nature, already implicit in the dichotomy of purpose and fact, is deeply embedded in the human consciousness and

[1] House of Commons, October 21, 1937, reprinted in N. Chamberlain, *The Struggle for Peace*, p. 33.

in political thought. The utopian sets up an ethical standard which purports to be independent of politics, and seeks to make politics conform to it. The realist cannot logically accept any standard value save that of fact. In his view, the absolute standard of the utopian is conditioned and dictated by the social order, and is therefore political. Morality can only be relative, not universal. Ethics must be interpreted in terms of politics ; and the search for an ethical norm outside politics is doomed to frustration. The identification of the supreme reality with the supreme good, which Christianity achieves by a bold stroke of dogmatism, is achieved by the realist through the assumption that there is no good other than the acceptance and understanding of reality.

These implications of the opposition between utopia and reality will emerge clearly from a more detailed study of the modern crisis in international politics.

PART TWO

THE INTERNATIONAL CRISIS

CHAPTER 3

THE UTOPIAN BACKGROUND

The Foundations of Utopianism

THE modern school of utopian political thought must be traced back to the break-up of the mediaeval system, which presupposed a universal ethic and a universal political system based on divine authority. The realists of the Renaissance made the first determined onslaught on the primacy of ethics and propounded a view of politics which made ethics an instrument of politics, the authority of the state being thus substituted for the authority of the church as the arbiter of morality. The answer of the utopian school to this challenge was not an easy one. An ethical standard was required which would be independent of any external authority, ecclesiastical or civil ; and the solution was found in the doctrine of a secular " law of nature " whose ultimate source was the individual human reason. Natural law, as first propounded by the Greeks, had been an intuition of the human heart about what is morally right. " It is eternal ", said Sophocles' Antigone, " and no man knows whence it came." The Stoics and the mediaeval schoolmen identified natural law with reason ; and in the seventeenth and eighteenth centuries this identification was revived in a new and special form. In science, the laws of nature were deduced by a process of reasoning from observed facts about the nature of matter. By an easy analogy, the Newtonian principles were now applied to the ethical problems. The moral law of nature could be scientifically established ; and rational deduction from the supposed facts of human nature took the place of revelation or intuition as the source of morality. Reason could determine what were the universally valid moral laws ; and the assumption was made that, once these laws were determined, human beings would conform to them just as matter conformed to

22

the physical laws of nature. Enlightenment was the royal road to the millennium.

By the eighteenth century, the main lines of modern utopian thought were firmly established. It was essentially individualist in that it made the human conscience the final court of appeal in moral questions ; in France it became associated with a secular, in England with an evangelical tradition. It was essentially rationalist in that it identified the human conscience with the voice of reason.[1] But it had still to undergo important developments ; and it was Jeremy Bentham who, when the industrial revolution had transferred the leadership of thought from France to England, gave to nineteenth-century utopianism its characteristic shape. Starting from the postulate that the fundamental characteristic of human nature is to seek pleasure and avoid pain, Bentham deduced from this postulate a rational ethic which defined the good in the famous formula " the greatest happiness of the greatest number ". As has often been pointed out, " the greatest happiness of the greatest number " performed the function, which natural law had performed for a previous generation, of an absolute ethical standard. Bentham firmly believed in this absolute standard, and rejected as " anarchical " the view that there are " as many standards of right and wrong as there are men ".[2] In effect, " the greatest happiness of the greatest number " was the nineteenth-century definition of the content of natural law.

The importance of Bentham's contribution was twofold. In the first place, by identifying the good with happiness, he provided a plausible confirmation of the " scientific " assumption of the eighteenth-century rationalists that man would infallibly conform to the moral law of nature once its content had been rationally determined. Secondly, while preserving the rationalist and individualist aspect of the doctrine, he succeeded in giving it a broader basis. The doctrine of reason in its eighteenth-century guise was pre-eminently intellectual and aristocratic. Its political corollary was an enlightened

---

[1] While this is the form of utopianism which has been predominant for the past three centuries, and which still prevails (though perhaps with diminishing force) in English-speaking countries, it would be rash to assert that individualism and rationalism are necessary attributes to utopian thought. Fascism contained elements of a utopianism which was anti-individualist and irrational. These qualities were already latent in the utopian aspects of Leninism — and perhaps even of Marxism.   [2] Bentham, *Works*, ed. Bowring, i. p. 31.

despotism of philosophers, who alone could be expected to have the necessary reasoning power to discover the good. But now that happiness was the criterion, the one thing needful was that the individual should understand where his happiness lay. Not only was the good ascertainable — as the eighteenth century had held — by a rational process, but this process — added the nineteenth century — was not a matter of abstruse philosophical speculation, but of simple common sense. Bentham was the first thinker to elaborate the doctrine of salvation by public opinion. The members of the community " may, in their aggregate capacity, be considered as constituting a sort of judicatory or tribunal — call it . . . *The Public-Opinion Tribunal* ".[1] It was James Mill, Bentham's pupil, who produced the most complete argument yet framed for the infallibility of public opinion :

> Every man possessed of reason is accustomed to weigh evidence and to be guided and determined by its preponderance. When various conclusions are, with their evidence, presented with equal care and with equal skill, there is a moral certainty, though some few may be misguided, that the greatest number will judge right, and that the greatest force of evidence, whatever it is, will produce the greatest impression.[2]

This is not the only argument by which democracy as a political institution can be defended. But this argument was, in fact, explicitly or implicitly accepted by most nineteenth-century liberals. The belief that public opinion can be relied on to judge rightly on any question rationally presented to it, combined with the assumption that it will act in accordance with this right judgment, is an essential foundation of the liberal creed. In Great Britain, the later eighteenth and the nineteenth centuries were pre-eminently the age of popular preaching and of political oratory. By the voice of reason men could be persuaded both to save their own immoral souls and to move along the path of political enlightenment and progress. The optimism of the nineteenth century was based on the triple conviction that the pursuit of the good was a matter of right reasoning, that the spread of knowledge would soon make it

---

[1] Bentham, *Works*, ed. Bowring, viii. p. 561.
[2] James Mill, *The Liberty of the Press*, pp. 22-3.

possible for everyone to reason rightly on this important subject, and that anyone who reasoned rightly on it would necessarily act rightly.

The application of these principles to international affairs followed, in the main, the same pattern. The Abbé Saint-Pierre, who propounded one of the earliest schemes for a League of Nations, " was so confident in the reasonableness of his projects that he always believed that, if they were fairly considered, the ruling powers could not fail to adopt them ".[1] Both Rousseau and Kant argued that, since wars were waged by princes in their own interest and not in that of their peoples, there would be no wars under a republican form of government. In this sense, they anticipated the view that public opinion, if allowed to make itself effective, would suffice to prevent war. In the nineteenth century, this view won widespread approval in Western Europe, and took on the specifically rationalist colour proper to the doctrine that the holding of the right moral beliefs and the performance of the right actions can be assured by process of reasoning. Never was there an age which so unreservedly proclaimed the supremacy of the intellect. " It is intellectual evolution ", averred Comte, " which essentially determines the main course of social phenomena." [2] Buckle, whose famous *History of Civilisation* was published between 1857 and 1861, boldly declared that dislike of war is " a cultivated taste peculiar to an intellectual people ". He chose a cogent example, based on the assumption, natural to a British thinker, of the ingrained bellicosity of Great Britain's most recent enemy. " Russia is a warlike country ", he wrote, " not because the inhabitants are immoral, but because they are unintellectual. The fault is in the head, not in the heart." [3] The view that the spread of education would lead to international peace was shared by many of Buckle's contemporaries and successors. Its last serious exponent was Sir Norman Angell, who sought, by *The Great Illusion* and other books, to convince the world that war never brought profit to anyone. If he could establish this point by irrefutable argument, thought Sir Norman, then war could not occur. War was simply a " failure of understanding ". Once the head was purged of

[1] J. S. Bury, *The Idea of Progress*, p. 131.
[2] Comte, *Cours de Philosophie Positive*, Lecture LXI.
[3] Buckle, *History of Civilisation* (World's Classics ed.), i. pp. 151-2.

the illusion that war was profitable, the heart could look after itself. " The world of the Crusades and of heretic burning ", ran the opening manifesto of a monthly journal called *War and Peace* which started publication in October 1913, ". . . was not a badly-meaning, but a badly-thinking world. . . . We emerged from it by correcting a defect in understanding ; we shall emerge from the world of political warfare or armed peace in the same way." [1] Reason could demonstrate the absurdity of the international anarchy ; and with increasing knowledge, enough people would be rationally convinced of its absurdity to put an end to it.

Benthamism Transplanted

Before the end of the nineteenth century, serious doubts had been thrown from more than one quarter on the assumptions of Benthamite rationalism. The belief in the sufficiency of reason to promote right conduct was challenged by psychologists. The identification of virtue with enlightened self-interest began to shock philosophers. The belief in the infallibility of public opinion had been attractive on the hypothesis of the earlier utilitarians that public opinion was the opinion of educated and enlightened men. It was less attractive, at any rate to those who thought themselves educated and enlightened, now that public opinion was the opinion of the masses ; and as early as 1859, in his essay *On Liberty*, J. S. Mill had been preoccupied with the dangers of " the tyranny of the majority ". After 1900, it would have been difficult to find, either in Great Britain or in any other European country, any serious political thinker who accepted the Benthamite assumptions without qualification. Yet, by one of the ironies of history, these half-discarded nineteenth-century assumptions reappeared, in the second and third decades of the twentieth century, in the special field of international politics, and there became the foundation-stones of a new utopian edifice. The explanation may be in part that, after 1914, men's minds naturally fumbled their way back, in search of a new utopia, to those apparently firm foundations of nineteenth-century peace and security. But a more decisive factor was the influence of the United States,

[1] Quoted in Angell, *Foundations of International Polity*, p. 224. Internal evidence suggests that the passage was written by Sir Norman Angell himself.

still in the heyday of Victorian prosperity and of Victorian belief in the comfortable Benthamite creed. Just as Bentham, a century earlier, had taken the eighteenth-century doctrine of reason and refashioned it to the needs of the coming age, so now Woodrow Wilson, the impassioned admirer of Bright and Gladstone, transplanted the nineteenth-century rationalist faith to the almost virgin soil of international politics and, bringing it back with him to Europe, gave it a new lease of life. Nearly all popular theories of international politics between the two world wars were reflexions, seen in an American mirror, of nineteenth-century liberal thought.

In a limited number of countries, nineteenth-century liberal democracy had been a brilliant success. It was a success because its presuppositions coincided with the stage of development reached by the countries concerned. Out of the mass of current speculation, the leading spirits of the age took precisely that body of theory which corresponded to their needs, consciously and unconsciously fitting their practice to it, and it to their practice. Utilitarianism and *laissez-faire* served, and in turn directed, the course of industrial and commercial expansion. But the view that nineteenth-century liberal democracy was based, not on a balance of forces peculiar to the economic development of the period and the countries concerned, but on certain *a priori* rational principles which had only to be applied in other contexts to produce similar results, was essentially utopian ; and it was this view which, under Wilson's inspiration, dominated the world after the first world war. When the theories of liberal democracy were transplanted, by a purely intellectual process, to a period and to countries whose stage of development and whose practical needs were utterly different from those of Western Europe in the nineteenth century, sterility and disillusionment were the inevitable sequel. Rationalism can create a utopia, but cannot make it real. The liberal democracies scattered throughout the world by the peace settlement of 1919 were the product of abstract theory, stuck no roots in the soil, and quickly shrivelled away.

Rationalism and the League of Nations

The most important of all the institutions affected by this one-sided intellectualism of international politics was the League of Nations, which was an attempt " to apply the principles of Lockeian liberalism to the building of a machinery of international order ".[1] " The Covenant ", observed General Smuts, ". . . simply carries into world affairs that outlook of a liberal democratic society which is one of the great achievements of our human advance." [2] But this transplantation of democratic rationalism from the national to the international sphere was full of unforeseen difficulties. The empiricist treats the concrete case on its individual merits. The rationalist refers it to an abstract general principle. Any social order implies a large measure of standardisation, and therefore of abstraction ; there cannot be a different rule for every member of the community. Such standardisation is comparatively easy in a community of several million anonymous individuals conforming more or less closely to recognised types. But it presents infinite complications when applied to sixty known states differing widely in size, in power, and in political, economic and cultural development. The League of Nations, being the first large-scale attempt to standardise international political problems on a rational basis, was particularly liable to these embarrassments.

The founders of the League, some of whom were men of political experience and political understanding, had indeed recognised the dangers of abstract perfection. " Acceptance of the political facts of the present ", remarked the official British Commentary on the Covenant issued in 1919, " has been one of the principles on which the Commission has worked ",[3] and this attempt to take account of political realities distinguished the Covenant not only from previous paper schemes of world organisation, but also from such purely utopian projects as the International Police Force, the Briand-Kellogg Pact and the United States of Europe. The Covenant possessed the virtue of several theoretical imperfections. Purporting to treat all members as equal, it assured to the Great

[1] R. H. S. Crossman in J. P. Mayer, *Political Thought*, p. 202.

[2] New Year's Eve broadcast from Radio-Nations, Geneva : *The Times*, January 1, 1938.

[3] *The Covenant of the League of Nations and a Commentary Thereon*, Cmd. 151 (1919), p. 12. " The great strength of the Covenant ", said the British Government some years later, " lies in the measure of discretion which it allows to the Council and Assembly in dealing with future contingencies which may have no parallel in history and which therefore cannot all of them be foreseen in advance " (*League of Nations : Official Journal*, May 1928, p. 703).

Powers a permanent majority on the Council of the League.[1]
It did not purport to prohibit war altogether, but only to limit
the occasions on which it might legitimately be resorted to.
The obligation imposed on members of the League to apply
sanctions to the Covenant-breaker was not free from vagueness ;
and this vagueness had been discreetly enhanced by a set of
" interpretative " resolutions passed by the Assembly of 1921.
The starkness of the territorial guarantee provided by Article 10
of the Covenant was smoothed away in a resolution which
secured an almost unanimous vote at the Assembly of 1923.
It seemed for the moment as if the League might reach a
working compromise between utopia and reality and become an
effective instrument of international politics.

Unhappily, the most influential European politicians neglected
the League during its critical formative years. Abstract rationalism gained the upper hand, and from about 1922 onwards the
current at Geneva set strongly in the utopian direction.[2] It
came to be believed, in the words of an acute critic, " that
there can exist, either at Geneva or in foreign offices, a sort of
carefully classified card-index of events or, better still, ' situations ', and that, when the event happens or the situation presents itself, a member of the Council or Foreign Minister can
easily recognise that event or situation and turn up the index
to be directed to the files where the appropriate action is pre-
scribed ".[3] There were determined efforts to perfect the
machinery, to standardise the procedure, to close the " gaps "
in the Covenant by an absolute veto on all war, and to make
the application of sanctions " automatic ". The Draft Treaty
of Mutual Assistance, the Geneva Protocol, the General Act,

[1] The defection of the United States upset this balance, and left four major
confronted with four minor Powers. Subsequent increases in membership, which
have taken place at frequent intervals since 1923, gave a permanent preponderance
to the minor Powers. The Council, in becoming more " representative ", lost
much of its effectiveness as a political instrument. Reality was sacrificed to an
abstract principle. It should be added that the prudent Swiss Delegate foresaw
this result when the first increase was mooted in 1922 (*League of Nations : Third
Assembly*, First Committee, pp. 37-8).

[2] By a curious irony, this development was strongly encouraged by a group of
American intellectuals ; and some European enthusiasts imagined that, by following this course, they would propitiate American opinion. The rift between the
theory of the intellectuals and the practice of the government, which developed in
Great Britain from 1932 onwards, began in the United States in 1919.

[3] J. Fischer-Williams, *Some Aspects of the Covenant of the League of Nations*,
p. 238.

the plan to incorporate the Briand-Kellogg Pact in the Covenant and " the definition of the aggressor ", were all milestones on the dangerous path of rationalisation. The fact that the utopian dishes prepared during these years at Geneva proved unpalatable to most of the principal governments concerned was a symptom of the growing divorce between theory and practice.

Even the language current in League circles betrayed the growing eagerness to avoid the concrete in favour of the abstract generalisations. When it was desired to arrange that the Draft Treaty of Mutual Assistance could be brought into force in Europe without waiting for the rest of the world, a stipulation was inserted that it might come into force " by continents " — a proviso with farcical implications for every continent except Europe. A conventional phraseology came into use, which served as the current coin of delegates at Geneva and of League enthusiasts elsewhere and which, through constant repetition, soon lost all contact with reality. " I cannot recall any time ", said Mr. Churchill in 1932, " when the gap between the kind of words which statesmen used and what was actually happening in many countries was so great as it is now." [1] The Franco-Soviet Pact, which was a defensive alliance against Germany, was so drafted as to make it appear an instrument of general application, and was described as a shining example of the principle of " collective security ". A member of the House of Commons, when asked in the debate on sanctions in June 1936 whether he would run the risk of war with Italy, replied that he was prepared to face " all the consequences naturally flowing from the enforcement of the Covenant against an aggressor nation ".[2] These linguistic contortions encouraged the frequent failure to distinguish between the world of abstract reason and the world of political reality. " Metaphysicians, like savages ", remarks Mr. Bertrand Russell, " are apt to imagine a magical connexion between words and things." [3] The metaphysicians of Geneva found it difficult to believe that an accumulation of ingenious texts prohibiting war was not a barrier against war itself. " Our purpose ", said M. Benes in introducing the Geneva Protocol to the 1924 Assembly, " was to make war impossible, to kill it, to anni-

[1] Winston Churchill, *Arms and the Covenant*, p. 43.
[2] Quoted in Toynbee, *Survey of International Affairs, 1935*, ii. p. 448.
[3] B. Russell in *Atlantic Monthly*, clix. (February 1937), p. 155.

hilate it. To do this we had to create a system." [1] The Protocol was the " system ". Such presumption could only provoke nemesis. Once it came to be believed in League circles that salvation could be found in a perfect card-index, and that the unruly flow of international politics could be canalised into a set of logically impregnable abstract formulae inspired by the doctrines of nineteenth-century liberal democracy, the end of the League as an effective political instrument was in sight.

The Apotheosis of Public Opinion

Nor did any better fortune attend the attempt to transplant to the international sphere the liberal democratic faith in public opinion. And here there was a double fallacy. The nineteenth-century belief in public opinion comprised two articles : first (and in democracies this was, with some reservations, true), that public opinion is bound in the long run to prevail ; and second (this was the Benthamite view), that public opinion is always right. Both these beliefs, not always clearly distinguished one from the other, were uncritically reproduced in the sphere of international politics.

The first attempts to invoke public opinion as a force in the international world had been made in the United States. In 1909, President Taft evolved a plan for the conclusion of treaties between the United States and other Great Powers for the compulsory arbitration of international disputes. But how, it was asked, would the award of the arbitral court be enforced ? Taft disposed of the question with complete light-heartedness. He had never observed that in a democracy like the United States the enforcement of awards gave rise to any particular difficulty ; and he professed himself " very little concerned " about this aspect of the matter. " After we have gotten the cases into court and decided, and the judgments embodied in a solemn declaration of a court thus established, few nations will care to face the condemnation of international public opinion and disobey the judgment." [2] Public opinion, as in democratic countries, was bound to prevail ; and public opinion, as the Benthamites said, could always be trusted to come down on the right side. The United States Senate rejected the President's proposal, so that the opportunity did not occur to

[1] *League of Nations : Fifth Assembly*, p. 497.
[2] W. Taft, *The United States and Peace*, p. 150.

put " international public opinion " to the test. Four years later, Bryan, Wilson's first Secretary of State, came forward with a further set of treaties. In the Bryan treaties, arbitration was dropped in favour of conciliation. Their most novel and significant feature was the provision that the parties to them should not resort to war until twelve months had elapsed from the beginning of the dispute. In hot blood, the Bryan treaties seemed to admit, men might not listen to the voice of reason. But once delay had cooled their passions, reason, in the guise of international public opinion, would resume her compelling force. Several such treaties were in fact signed between the United States and other Powers — some of them, by a curious irony, in the first days of the first world war. " The sum and substance " of these treaties, said Wilson in October 1914, was " that whenever any trouble arises the light shall shine on it for a year before anything is done ; and my prediction is that after the light has shone on it for a year, it will not be necessary to do anything ; that after we know what happened, then we will know who was right and who was wrong ".[1]

The belief in the compelling power of reason, expressed through the voice of the people, was particularly congenial to Wilson. When he entered politics in 1910 as a candidate for the Governorship of New Jersey, his campaign was based on an appeal to " the people " against the political bosses ; and he displayed " an almost mystical faith that the people would follow him if he could speak to enough of them ". The result of his campaign confirmed him in his belief in the potency of the voice of reason speaking through his lips. He would govern by the persuasiveness of reason acting on an all-powerful public opinion. " If the bosses held back, he had only to appeal to the people. . . . The people wanted the high things, the right things, the true things." [2]

America's entry into the war entailed no modification of Wilson's faith in the rightness of popular judgment. He took up the cue in one of the speeches in which he discussed the future conditions of peace :

> It is the peculiarity of this great war that, while statesmen have seemed to cast about for definitions of their purpose

[1] *The Public Papers of Woodrow Wilson : The New Democracy*, ed. R. S. Baker, i. p. 206.
[2] R. S. Baker, *Woodrow Wilson : Life and Letters*, iii. p. 173.

and have sometimes seemed to shift their ground and their point of view, the thought of the mass of men, whom statesmen are supposed to instruct and lead, has grown more and more unclouded, more and more certain of what it is they are fighting for. National purposes have fallen more and more into the background ; and the common purpose of enlightened mankind has taken their place. The counsels of plain men have become on all hands more simple and straightforward and more unified than the counsels of sophisticated men of affairs, who still retain the impression that they are playing a game of power and are playing for high stakes. That is why I have said that this is a people's war, not a statesmen's. Statesmen must follow the clarified common thought or be broken.[1]

" Unless the Conference was prepared to follow the opinions of mankind ", he said on his way to Paris, " and to express the will of the people rather than that of the leaders of the Conference, we should be involved in another break-up of the world." [2]

Such conceptions did, in fact, play a conspicuous part in the work of the Conference. When the Italian Delegates proved recalcitrant in their claims to Fiume and the Adriatic coast, Wilson remained convinced that if he could appeal against the " leaders " to the " people ", if only (as at the New Jersey election) he " could speak to enough of them ", the voice of reason must infallibly prevail. The communiqué to the Italian people, and the withdrawal of the Italian Delegation from Paris, were the result of this conviction. The problem of disarmament was approached in the same spirit. Once the enemy Powers had been disarmed by force, the voice of reason, speaking through public opinion, could be trusted to disarm the Allies. Both Wilson and Mr. Lloyd George " felt that, if the German army was limited, France would have to follow suit, and that she could hardly maintain an immense army under those conditions ".[3] And if anyone had paused to enquire on what compulsion France would have to disarm, the only answer could have been an appeal to the rational force of public

---

[1] *The Public Papers of Woodrow Wilson : War and Peace*, ed. R. S. Baker, i. p. 259.

[2] *Intimate Papers of Colonel House*, ed. C. Seymour, iv. p. 291.

[3] D. Lloyd George, *The Truth about the Treaties*, i. p. 187.

opinion. Most important of all, the whole conception of the League of Nations was from the first closely bound up with the twin belief that public opinion was bound to prevail and that public opinion was the voice of reason. If "open covenants openly arrived at" could be made the rule, the plain people could be relied on to see that the contents conformed to the requirements of that reason which was the highest morality. The new order must be based, not on "covenants of selfishness and compromise" between governments, but on "the thought of the plain people here and everywhere throughout the world, the people who enjoy no privilege and have very simple and unsophisticated standards of right and wrong ".[1] It must be " sustained by the organised opinion of mankind ".[2]

The ticklish problem of material sanctions was approached reluctantly from the American, and almost as reluctantly from the British, side. Like Taft, Anglo-Saxon opinion felt itself " very little concerned " over this aspect of the matter ; for the recognition of the necessity of sanctions was in itself a derogation from the utopian doctrine of the efficacy of rational public opinion. It was unthinkable that a unanimous verdict of the League should be defied ; and even if by some mischance the verdict were not unanimous, " a majority report would probably be issued, and . . . this ", suggested Lord Cecil during the debates in Paris, " would be likely to carry great weight with the public opinion of the world ".[3] The official British Commentary on the Covenant developed the same point of view :

> The League [it declared] must continue to depend on the free consent, in the last resort, of its component States ; this assumption is evident in nearly every article of the Covenant, of which the ultimate and most effective sanction must be the public opinion of the civilised world. If the nations of the future are in the main selfish, grasping and warlike, no instrument or machinery will restrain them. It is only possible to establish an organisation which may make peaceful co-operation easy and hence customary, and to trust in the influence of custom to mould public opinion.

[1] *The Public Papers of Woodrow Wilson : War and Peace*, ed. R. S. Baker, i. p. 133.

[2] *Ibid.* i. p. 234.

[3] Miller, *The Drafting of the Covenant*, ii. p. 64.

The sanctions provisions were slurred over, half apologetically and with a consolatory postscript :

> Not the least important part of the pressure will be supplied by the publicity stipulated for in the procedure of settlement. The obscure issues from which international quarrels arise will be dragged out into the light of day and the creation of an informed public opinion made possible.[1]

When the House of Commons debated the ratification of the Versailles Treaty, Lord Cecil was the principal expositor of the League Covenant :

> For the most part [he told the House] there is no attempt to rely on anything like a superstate ; no attempt to rely upon force to carry out a decision of the Council or the Assembly of the League. That is almost impracticable as things stand now. What we rely upon is public opinion . . . and if we are wrong about it, then the whole thing is wrong.[2]

Addressing the Imperial Conference of 1923 on the subject of the League, Lord Cecil explained that " its method is not . . . the method of coercive government : it is a method of consent and its executive instrument is not force, but public opinion ".[3] And when the first League Assembly met, Lord Cecil, as British Delegate, propounded the same philosophy from the tribune :

> It is quite true that by far the most powerful weapon at the command of the League of Nations is not the economic or the military weapon or any other weapon of material force. By far the strongest weapon we have is the weapon of public opinion.[4]

Even the more sceptical and sophisticated Balfour, explaining the absence of sanctions from the Washington agreements of 1921, declared that " if any nation hereafter deliberately separates itself from the collective action we have taken in Washington in this year of grace, it will stand condemned

---

[1] *The Covenant of the League of Nations with a Commentary Thereon*, Cmd. 151, pp. 12, 16.

[2] House of Commons, July 21, 1919 : *Official Report*, cols. 990, 992.

[3] *Imperial Conference of 1923*, Cmd. 1987, p. 44.

[4] *League of Nations : First Assembly*, p. 395.

before the world " ; [1] and it was one of the presuppositions of liberal democracy that such condemnation would be effective. But the argument that public opinion is the all-important weapon is two-edged ; and in 1932, during the Manchurian crisis, the ingenious Sir John Simon used it to demonstrate that any other kind of action was superfluous. " The truth is ", he told the House of Commons, " that when public opinion, world opinion, is sufficiently unanimous to pronounce a firm moral condemnation, sanctions are not needed." [2] Given the Benthamite and Wilsonian premises, this answer was irrefutable. If public opinion had failed to curb Japan, then — as Lord Cecil had said in 1919 — " the whole thing is wrong ".

The Nemesis of Utopianism

The nemesis of utopianism in international politics came rather suddenly. In September 1930, the President of Columbia University, Dr. Nicholas Murray Butler, ventured on the " reasonably safe prediction that the next generation will see a constantly increasing respect for Cobden's principles and point of view and a steadily growing endeavour more largely to give them practical effect in public policy ".[3] On September 10, 1931, Lord Cecil told the Assembly of the League of Nations that " there has scarcely ever been a period in the world's history when war seems less likely than it does at present ".[4] On September 18, 1931, Japan opened her campaign in Manchuria ; and in the following month, the last important country which had continued to adhere to the principle of free trade took the first steps towards the introduction of a general tariff.

From this point onwards, a rapid succession of events forced upon all serious thinkers a reconsideration of premises which were becoming more and more flagrantly divorced from reality. The Manchurian crisis had demonstrated that the " condemnation of international public opinion ", invoked by Taft and by so many after him, was a broken reed. In the United States, this conclusion was drawn with extreme reluctance. In 1932, an American Secretary of State still cautiously main-

[1] Quoted in Zimmern, *The League of Nations and the Rule of Law*, p. 399.
[2] House of Commons, March 22, 1932 : *Official Report*, col. 923.
[3] N. M. Butler, *The Path to Peace*, p. xii.
[4] *League of Nations : Twelfth Assembly*, p. 59.

tained that " the sanction of public opinion can be made one of the most potent sanctions of the world ".[1] In September 1938, President Roosevelt based his intervention in the Czecho-Slovak crisis on the belief of the United States Government in " the moral force of public opinion " ; [2] and in April 1939, Mr. Cordell Hull once again announced the conviction that " a public opinion, the most potent of all forces for peace, is more strongly developing throughout the world ".[3] But in countries more directly menaced by international crisis, this consoling view no longer found many adherents ; and the continued addition to it of American statesmen was regarded as an index of American unwillingness to resort to more potent weapons. Already in 1932, Mr. Churchill taunted the League of Nations Union with " long-suffering and inexhaustible gullibility " for continuing to preach this outworn creed.[4] Before long the group of intellectuals who had once stressed the relative unimportance of the " material " weapons of the League began to insist loudly on economic and military sanctions as the necessary cornerstones of an international order. When Germany annexed Austria, Lord Cecil indignantly enquired whether the Prime Minister " holds that the use of material force is impracticable and that the League should cease to attempt ' sanctions ' and confine its efforts to moral force ".[5] The answer might well have been that, if Neville Chamberlain did in fact hold this view, he could have learned it from Lord Cecil's own earlier utterances.

Moreover, scepticism attacked not only the premise that public opinion is certain to prevail, but also the premise that public opinion is certain to be right. At the Peace Conference, it had been observed that statesmen were sometimes more reasonable and moderate in their demands than the public opinion which they were supposed to represent. Even Wilson himself once used — no doubt, in perfect sincerity — an argument which directly contradicted his customary thesis that reason can be made to prevail by appealing to " the plain

---

[1] Mr. Stimson to the Council of Foreign Relations on August 8, 1932 (*New York Times*, August 9, 1932).

[2] " Believing, as this government does, in the moral force of public opinion . . ." (Sumner Welles in *State Department Press Releases*, October 8, 1938, p. 237).

[3] *The Times*, April 18, 1939.

[4] Winston Churchill, *Arms and the Covenant*, p. 36.

[5] *Daily Telegraph*, March 24, 1938.

people everywhere throughout the world ". In the League of Nations Commission of the Conference, the Japanese had raised the issue of race equality. " How can you treat on its merits in this quiet room ", enquired the President, " a question which will not be treated on its merits when it gets out of this room ? " [1] Later history provided many examples of this phenomenon. It became a commonplace for statesmen at Geneva and elsewhere to explain that they themselves had every desire to be reasonable, but that public opinion in their countries was inexorable ; and though this plea was sometimes a pretext or a tactical manœuvre, there was often a solid substratum of reality beneath it. The prestige of public opinion correspondingly declined. " It does not help the conciliator, the arbitrator, the policeman or the judge ", wrote a well-known supporter of the League of Nations Union recently, " to be surrounded by a crowd emitting either angry or exulting cheers." [2] Woodrow Wilson's " plain men throughout the world ", the spokesmen of " the common purpose of enlightened mankind ", had somehow transformed themselves into a disorderly mob emitting incoherent and unhelpful noises. It seemed undeniable that, in international affairs, public opinion was almost as often wrong-headed as it was impotent. But where so many of the presuppositions of 1919 were crumbling, the intellectual leaders of the utopian school stuck to their guns ; and in Great Britain and the United States — and to a lesser degree in France — the rift between theory and practice assumed alarming dimensions. Armchair students of international affairs were unanimous about the kind of policy which ought to be followed, both in the political and in the economic field. Governments of many countries acted in a sense precisely contrary to this advice, and received the endorsement of public opinion at the polls.

The Problem of Diagnosis

In such disasters the obvious explanation is never far to seek. The able historian of the Communist International has noted that, in the history of that institution, " every failure — not objective failure, but the failure of the reality to comply

---

[1] Miller, *The Drafting of the Covenant*, ii. p. 701.
[2] Lord Allen of Hurtwood, *The Times*, May 30, 1938.

with the utopia — supposes a traitor ".[1]   The principle has a wide application, and touches deep springs of human character. Statesmen of more than one country have been pilloried by disappointed utopians as wreckers of the international order. The few members of the school who have tried to go behind this simple anthropomorphic explanation hesitate between two alternative diagnoses.   If mankind in its international relations has signally failed to achieve the rational good, it must either have been too stupid to understand that good, or too wicked to pursue it.   Professor Zimmern leans to the hypothesis of stupidity, repeating almost word for word the argument of Buckle and Sir Norman Angell :

> The obstacle in our path . . . is not in the moral sphere, but in the intellectual. . . . It is not because men are ill-disposed that they cannot be educated into a world social consciousness.   It is because they — let us be honest and say " we " — are beings of conservative temper and limited intelligence.

The attempt to build a world order has failed not through " pride or ambition or greed ", but through " muddled thinking ".[2]   Professor Toynbee, on the other hand, sees the cause of the breakdown in human wickedness.   In a single volume of the annual *Survey of International Affairs*, he accuses Italy of " positive, strong-willed, aggressive egotism ", Great Britain and France of " negative, weak-willed, cowardly egotism ", Western Christendom as a whole of a " sordid " crime, and all the members of the League of Nations, except Abyssinia, of "covetousness " or "cowardice " (the choice is left to them), while the attitude of the Americans is merely " rather captious and perverse ".[3]   Some writers combined the charge of stupidity and the charge of wickedness.   Much comment on international affairs was rendered tedious and sterile by incessant girding at a reality which refused to conform to utopian prescriptions.

The simplicity of these explanations seemed almost ludicrously disproportionate to the intensity and complexity of the

---

[1] F. Borkenau, *The Communist International*, p. 179.

[2] *Neutrality and Collective Security* (Harris Foundation Lectures : Chicago, 1936), pp. 8, 18.

[3] *Survey of International Affairs, 1935*, ii. pp. 2, 89, 96, 219-20, 480.

international crisis.[1] The impression made on the ordinary man was more accurately recorded in April 1938 in some words of Mr. Anthony Eden :

> It is utterly futile to imagine that we are involved in a European crisis which may pass as it has come. We are involved in a crisis of humanity all the world over. We are living in one of those great periods of history which are awe-inspiring in their responsibilities and in their consequences. Stupendous forces are loose, hurricane forces.[2]

It is not true, as Professor Toynbee believes, that we have been living in an exceptionally wicked age. It is not true, as Professor Zimmern implies, that we have been living in an exceptionally stupid one. Still less is it true, as Professor Lauterpacht more optimistically suggests, that what we have been experiencing is " a transient period of retrogression " which should not be allowed unduly to colour our thought.[3] It is a meaningless evasion to pretend that we have witnessed, not the failure of the League of Nations, but only the failure of those who refused to make it work. The breakdown of the nineteen-thirties was too overwhelming to be explained merely in terms of individual action or inaction. Its downfall involved the bankruptcy of the postulates on which it was based. The foundations of nineteenth-century belief are themselves under suspicion. It may be not that men stupidly or wickedly failed to apply right principles, but that the principles themselves were false or inapplicable. It may turn out to be untrue that if men reason rightly about international politics they will also act rightly, or that right reasoning about one's own or one's nation's interests is the road to an international paradise. If the assumptions of nineteenth-century liberalism are in fact untenable, it need not surprise us that the utopia of the international theorists made so little impression on reality. But if they are untenable to-day, we shall also have to explain why they found such widespread acceptance, and inspired such splendid achievements, in the nineteenth century.

[1] As a recent writer has said of the French eighteenth-century rationalists, " their superficiality lay in a shocking exaggeration of the simplicity of the problem " Sabine, *A History of Political Theory*, p. 551).

[2] Anthony Eden, *Foreign Affairs*, p. 275.

[3] *International Affairs*, xvii. (September–October 1938), p. 712.

CHAPTER 4

THE HARMONY OF INTERESTS

The Utopian Synthesis

No political society, national or international, can exist unless people submit to certain rules of conduct. The problem why people should submit to such rules is the fundamental problem of political philosophy. The problem presents itself just as insistently in a democracy as under other forms of government and in international as in national politics ; for such a formula as " the greatest good of the greatest number " provides no answer to the question why the minority, whose greatest good is *ex hypothesi* not pursued, should submit to rules made in the interest of the greatest number. Broadly speaking, the answers given to the question fall into two categories, corresponding to the antithesis, discussed in a previous chapter, between those who regard politics as a function of ethics and those who regard ethics as a function of politics.

Those who assert the primacy of ethics over politics will hold that it is the duty of the individual to submit for the sake of the community as a whole, sacrificing his own interest to the interest of others who are more numerous, or in some other way more deserving. The " good " which consists in self-interest should be subordinated to the " good " which consists in loyalty and self-sacrifice for an end higher than self-interest. The obligation rests on some kind of intuition of what is right and cannot be demonstrated by rational argument. Those, on the other hand, who assert the primacy of politics over ethics, will argue that the ruler rules because he is the stronger, and the ruled submit because they are the weaker. This principle is just as easily applicable to democracy as to any other form of government. The majority rules because it is stronger, the minority submits because it is weaker. Democracy, it has often been said, substitutes the counting of heads for the breaking of heads. But the substitution is merely a convenience, and the principle of the two methods is the same. The realist, therefore, unlike the intuitionist, has a perfectly rational answer to the question why the individual should submit. He should submit because otherwise the stronger will compel him ; and the results of compulsion are more disagreeable than those of voluntary submission. Obligation is thus derived from a sort of spurious ethic based on the reasonableness of recognising that might is right.

Both these answers are open to objection. Modern man, who has witnessed so many magnificent achievements of human reason, is reluctant to believe that reason and obligation sometimes conflict. On the other hand, men of all ages have failed to find satisfaction in the view that the rational basis of obligation is merely the right of the stronger. One of the strongest points of eighteenth- and nineteenth-century utopianism was its apparent success in meeting both these objections at once. The utopian, starting from the primacy of ethics, necessarily believes in an obligation which is ethical in character and independent of the right of the stronger. But he has also been able to convince himself, on grounds other than those of the realist, that the duty of the individual to submit to rules made in the interest of the community can be justified in terms of reason, and that the greatest good of the greatest number is a rational end even for those who are not included in the greatest number. He achieves this synthesis by maintaining that the highest interest of the individual and the highest interest of the community naturally coincide. In pursuing his own interest, the individual pursues that of the community, and in promoting the interest of the community he promotes his own. This is the famous doctrine of the harmony of interests. It is a necessary corollary of the postulate that moral laws can be established by right reasoning. The admission of any ultimate divergence of interests would be fatal to this postulate ; and any apparent clash of interests must therefore be explained as the result of wrong calculation. Burke tacitly accepted the doctrine of identity when he defined expediency as " that which is good for the community and for every individual in it ".[1] It was handed on from the eighteenth-century rationalists to Bentham, and from Bentham to the Victorian moralists. The utilitarian philosophers could justify morality by the argument that, in promoting the good of others, one automatically promotes one's own. Honesty is the best policy. If people or nations behave

[1] Burke, *Works*, v. 407.

badly, it must be, as Buckle and Sir Norman Angell and Professor Zimmern think, because they are unintellectual and short-sighted and muddle-headed.

The Paradise of Laissez-Faire

It was the *laissez-faire* school of political economy created by Adam Smith which was in the main responsible for popularising the doctrine of the harmony of interests. The purpose of the school was to promote the removal of state control in economic matters ; and in order to justify this policy, it set out to demonstrate that the individual could be relied on, without external control, to promote the interests of the community for the very reason that those interests were identical with his own. This proof was the burden of *The Wealth of Nations*. The community is divided into those who live by rent, those who live by wages and those who live by profit ; and the interests of " those three great orders " are " strictly and inseparably connected with the general interest of the society ".[1] The harmony is none the less real if those concerned are unconscious of it. The individual " neither intends to promote the public interest, nor knows how much he is promoting it. . . . He intends only his own gain, and he is in this, as in many other cases, led by an invisible hand to promote an end which was no part of his intention." [2] The invisible hand, which Adam Smith would perhaps have regarded as a metaphor, presented no difficulty to Victorian piety. " It is curious to observe ", remarks a tract issued by the Society for the Propagation of Christian Knowledge towards the middle of the nineteenth century, " how, through the wise and beneficent arrangement of Providence, men thus do the greatest service to the public when they are thinking of nothing but their own gain." [3] About the same time an English clergyman wrote a work entitled *The Temporal Benefits of Christianity Explained*. The harmony of interests provided a solid rational basis for morality. To love one's neighbour turned out to be a thoroughly enlightened way of loving oneself. " We now know ", wrote Mr. Henry Ford as recently as 1930, " that anything which is

[1] Adam Smith, *The Wealth of Nations*, Book I. ch. xi. conclusion.
[2] *Ibid.* Book IV. ch. ii.
[3] Quoted in J. M. Keynes, *A Tract on Monetary Reform*, p. 7.

economically right is also morally right. There can be no conflict between good economics and good morals." [1]

The assumption of a general and fundamental harmony of interests is *prima facie* so paradoxical that it requires careful scrutiny. In the form which Adam Smith gave to it, it had a definite application to the economic structure of the eighteenth century. It presupposed a society of small producers and merchants, interested in the maximisation of production and exchange, infinitely mobile and adaptable, and unconcerned with the problem of the distribution of wealth. Those conditions were substantially fulfilled in an age when production involved no high degree of specialisation and no sinking of capital in fixed equipment, and when the class which might be more interested in the equitable distribution of wealth than in its maximum production was insignificant and without influence. But by a curious coincidence, the year which saw the publication of *The Wealth of Nations* was also the year in which Watt invented his steam-engine. Thus, at the very moment when *laissez-faire* theory was receiving its classical exposition, its premises were undermined by an invention which was destined to call into being immobile, highly specialised, mammoth industries and a large and powerful proletariat more interested in distribution than in production. Once industrial capitalism and the class system had become the recognised structure of society, the doctrine of the harmony of interests acquired a new significance, and became, as we shall presently see, the ideology of a dominant group concerned to maintain its predominance by asserting the identity of its interests with those of the community as a whole. [2]

But this transformation could not have been effected, and the doctrine could not have survived at all, but for one circumstance. The survival of the belief in a harmony of interests was rendered possible by the unparalleled expansion of production, population and prosperity, which marked the hundred years following the publication of *The Wealth of Nations* and the invention of the steam-engine. Expanding prosperity contributed to the popularity of the doctrine in three different ways. It attenuated competition for markets among producers, since fresh markets were constantly becoming available ; it

---

[1] Quoted in J. Truslow Adams, *The Epic of America*, p. 400. I have failed to trace the original.   [2] See pp. 80-81.

postponed the class issue, with its insistence on the primary importance of equitable distribution, by extending to members of the less prosperous classes some share in the general prosperity ; and by creating a sense of confidence in present and future well-being, it encouraged men to believe that the world was ordered on so rational a plan as the natural harmony of interests. " It was the continual widening of the field of demand which, for half a century, made capitalism operate as if it were a liberal utopia." [1] The tacit presupposition of infinitely expanding markets was the foundation on which the supposed harmony of interests rested. As Dr. Mannheim points out, traffic control is unnecessary so long as the number of cars does not exceed the comfortable capacity of the road.[2] Until that moment arrives, it is easy to believe in a natural harmony of interests between road-users.

What was true of individuals was assumed to be also true of nations. Just as individuals, by pursuing their own good, unconsciously compass the good of the whole community, so nations in serving themselves serve humanity. Universal free trade was justified on the ground that the maximum economic interest of each nation was identified with the maximum economic interest of the whole world. Adam Smith, who was a practical reformer rather than a pure theorist, did indeed admit that governments might have to protect certain industries in the interests of national defence. But such derogations seemed to him and to his followers trivial exceptions to the rule. " *Laissez-faire* ", as J. S. Mill puts it, ". . . should be the general rule : every departure from it, unless required by some great good, a certain evil." [3] Other thinkers gave the doctrine of the harmony of national interests a still wider application. " The true interests of a nation ", observes a late eighteenth-century writer, " never yet stood in opposition to the general interest of mankind ; and it can never happen that philanthropy and patriotism can impose on any man inconsistent duties." [4] T. H. Green, the English Hegelian who tempered the doctrines of his master with concessions to British nineteenth-century liberalism, held that " no action in its own interest of a state

---

[1] *Nationalism: A Study by a Group of Members of the Royal Institute of International Affairs*, p. 229.

[2] K. Mannheim, *Mensch und Gesellschaft im Zeitalter des Umbaus*, p. 104.

[3] J. S. Mill, *Principles of Political Economy*, II. Book V. ch. xi.

[4] Romilly, *Thoughts on the Influence of the French Revolution*, p. 5.

which fulfilled its idea could conflict with any true interest or right of general society ",[1] though it is interesting to note that the question-begging epithet " true ", which in the eighteenth-century quotation is attached to the interests of the nation, has been transferred by the nineteenth century to the interest of the general society. Mazzini, who embodied the liberal nineteenth-century philosophy of nationalism, believed in a sort of division of labour between nations. Each nation had its own special task for which its special aptitudes fitted it, and the performance of this task was its contribution to the welfare of humanity. If all nations acted in this spirit, international harmony would prevail. The same condition of apparently infinite expansibility which encouraged belief in the economic harmony of interests made possible the belief in the political harmony of rival national movements. One reason why contemporaries of Mazzini thought nationalism a good thing was that there were few recognised nations, and plenty of room for them. In an age when Germans, Czechs, Poles, Ukrainians, Magyars and half a dozen more national groups were not yet visibly jostling one another over an area of a few hundred square miles, it was comparatively easy to believe that each nation, by developing its own nationalism, could make its own special contribution to the international harmony of interests. Most liberal writers continued to believe, right down to 1918, that nations, by developing their own nationalism, promoted the cause of internationalism ; and Wilson and many other makers of the peace treaties saw in national self-determination the key to world peace. More recently still, responsible Anglo-Saxon statesmen have been from time to time content to echo, probably without much reflexion, the old Mazzinian formulae.[2]

Darwinism in Politics

When the centenary of *The Wealth of Nations* was celebrated in 1876, there were already symptoms of an impending breakdown. No country but Great Britain had been commercially powerful enough to believe in the international harmony of

---

[1] T. H. Green, *Principles of Political Obligation*, § 166.

[2] Mr. Eden, for example, in 1938 advocated " a comity of nations in which each can develop and flourish and give to their uttermost their own special contribution to the diversity of life " (Anthony Eden, *Foreign Affairs*, p. 277).

economic interests. Acceptance of free-trade principles outside
Great Britain had always been partial, half-hearted and short-
lived   The United States had rejected them from the start.
About 1840, Friedrich List, who had spent much time studying
industrial development in the United States, began to preach
to a German audience the doctrine that, while free trade was
the right policy for an industrially dominant nation like Great
Britain, only protection could enable weaker nations to break
the British stranglehold.   German and American industries,
built up behind protective tariffs, were soon seriously impinging
on the world-wide British industrial monopoly.   The British
Dominions overseas made use of their newly-won fiscal auto-
nomy to protect themselves against the manufactures of the
mother country.   The pressure of competition was increasing
on all sides.   Nationalism began to wear a sinister aspect,
and to degenerate into imperialism.   The philosophy of Hegel,
who identified reality with an eternally recurring conflict of
ideas, extended its influence.   Behind Hegel stood Marx, who
materialised the Hegelian conflict into a class-war of economic
interest-groups, and working-class parties came into being
which steadfastly refused to believe in the harmony of interests
between capital and labour.   Above all, Darwin propounded
and popularised a biological doctrine of evolution through a
perpetual struggle for life and the elimination of the unfit.

It was the doctrine of evolution which for a time enabled
the *laissez-faire* philosophy to make its terms with the new
conditions and the new trend of thought.   Free competition
had always been worshipped as the beneficent deity of the
laissez-faire system.   The French economist Bastiat, in a work
significantly entitled *Les Harmonies Économiques*, had hailed
competition as " that humanitarian force . . . which continually wrests progress from the hands of the individual to make
it the common heritage of the great human family ".[1]   Under
the growing strains of the latter half of the nineteenth century,
it was perceived that competition in the economic sphere
implied exactly what Darwin proclaimed as the biological law
of nature — the survival of the stronger at the expense of the
weaker.   The small producer or trader was gradually being
put out of business by his large-scale competitor ; and this
development was what progress and the welfare of the com-

[1] Bastiat, *Les Harmonies Économiques*, p. 355.

munity as a whole demanded. *Laissez-faire* meant an open field, and the prize to the strongest. The doctrine of the harmony of interests underwent an almost imperceptible modification. The good of the community (or, as people were now inclined to say, of the species) was still identical with the good of its individual members, but only of those individuals who were effective competitors in the struggle for life. Humanity went on from strength to strength, shedding its weaklings by the way. " The development of the species ", as Marx said, ". . . and therefore the higher development of the individual, can only be secured through the historical process, in which individuals are sacrificed." [1] Such was the doctrine of the new age of intensified economic competition preached by the school of Herbert Spencer, and commonly accepted in Great Britain in the 'seventies and 'eighties. The last French disciple of Adam Smith, Yves Guyot, assisted perhaps by the accident that the French word *concurrence* means " collaboration " as well as " competition ", wrote a work entitled *La Morale de la Concurrence*. Among English writers who applied this evolutionary principle to international politics, the most popular was Bagehot :

> Conquest is the premium given by nature to those national characters which their national customs have made most fit to win in war, and in most material respects those winning characters are really the best characters. The characters which do win in war are the characters which we should wish to win in war.[2]

About the same time, a Russian sociologist defined international politics as " the art of conducting the struggle for existence between social organisms " ; [3] and in 1900 a distinguished professor, in a once famous book, stated the doctrine in all its naked ruthlessness :

> The path of progress is strewn with the wreck of nations ; traces are everywhere to be seen of the hecatombs of inferior races, and of victims who found not the narrow way to the

[1] Marx, *Theorien über den Mehrwert*, II. i. p. 309.
[2] Bagehot, *Physics and Politics* (2nd ed.), p. 215. What does " material " mean in this passage ? Does it merely mean " relevant " ? Or is the writer conscious of an uncomfortable antithesis between " material " and " moral " ?
[3] J. Novicow, *La Politique Internationale*, p. 242.

greater perfection. Yet these dead peoples are, in very truth, the stepping stones on which mankind has arisen to the higher intellectual and deeper emotional life of to-day.[1]

In Germany, the same view was propounded by Treitschke and Houston Stewart Chamberlain. The doctrine of progress through the elimination of unfit nations seemed a fair corollary of the doctrine of progress through the elimination of unfit individuals ; and some such belief, though not always openly avowed, was implicit in late nineteenth-century imperialism. In the later nineteenth century, as an American historian remarks, " the basic problem of international relations was who should cut up the victims ".[2] The harmony of interests was established through the sacrifice of " unfit " Africans and Asiatics.

One point had, unfortunately, been overlooked. For more than a hundred years, the doctrine of the harmony of interests had provided a rational basis for morality. The individual had been urged to serve the interest of the community on the plea that that interest was also his own. The ground had now been shifted. In the long run, the good of the community and the good of the individual were still the same. But this eventual harmony was preceded by a struggle for life between individuals, in which not only the good, but the very existence, of the loser were eliminated altogether from the picture. Morality in these conditions had no rational attraction for prospective losers ; and the whole ethical system was built on the sacrifice of the weaker brother. In practice, nearly every state had made inroads on the classical doctrine, and introduced social legislation to protect the economically weak against the economically strong. The doctrine itself died harder. In the 'seventies Dostoevsky, who had none of the prejudices of an Englishman or an economist, made Ivan Karamazov declare that the price of admission to the " eternal harmony " was too high if it included the sufferings of the innocent. About the same time, Winwood Reade made an uncomfortable sensation in Great Britain with a book called *The Martyrdom of Man*, which drew attention to the immense tale of suffering and waste involved in the theory of evolution. In the 'nineties, Huxley

[1] Karl Pearson, *National Life from the Standpoint of Science*, p. 64.
[2] W. L. Langer, *The Diplomacy of Imperialism*, ii. p. 797.

confessed, in the name of science, to the existence of a discrepancy between the " cosmic process " and the " ethical process " ;[1] and Balfour, approaching the problem from the angle of philosophy, concluded that " a complete harmony between ' egoism ' and ' altruism ', between the pursuit of the highest happiness for oneself and the highest happiness for other people, can never be provided by a creed which refuses to admit that the deeds done and the character formed in this life can flow over into another, and there permit a reconciliation and an adjustment between the conflicting principles which are not always possible here ".[2] Less and less was heard of the beneficent properties of free competition. Before 1914, though the policy of international free trade was still upheld by the British electorate and by British economists, the ethical postulate which had once formed the basis of the *laissez-faire* philosophy no longer appealed, at any rate in its crude form, to any serious thinker. Biologically and economically, the doctrine of the harmony of interests was tenable only if you left out of account the interest of the weak who must be driven to the wall, or called in the next world to redress the balance of the present.

The International Harmony

Attention has been drawn to the curious way in which doctrines, already obsolete or obsolescent before the war of 1914, were reintroduced in the post-war period, largely through American inspiration, into the special field of international affairs. This would appear to be conspicuously true of the *laissez-faire* doctrine of the harmony of interests. In the United States, the history of *laissez-faire* presents special features. Throughout the nineteenth, and well into the twentieth, centuries the United States, while requiring tariff protection against European competition, had enjoyed the advantage of an expanding domestic market of apparently unlimited potentialities. In Great Britain, which continued down to 1914 to dominate world trade, but was increasingly conscious of strains and stresses at home, J. S. Mill and later economists clung firmly to international free trade, but made more and more inroads into *laissez-faire* orthodoxy in the domestic sphere. In the

---

[1] Huxley, Romanes Lecture, 1893, reprinted in *Evolution and Ethics*, p. 81.

[2] Balfour, *Foundations of Belief*, p. 27.

United States, Carey and his successors justified protective tariffs, but in every other respect maintained the immutable principles of *laissez-faire*. In Europe after 1919, planned economy, which rests on the assumption that no natural harmony of interests exists and that interests must be artificially harmonised by state action, became the practice, if not the theory, of almost every state. In the United States, the persistence of an expanding domestic market staved off this development till after 1929. The natural harmony of interests remained an integral part of the American view of life ; and in this as in other respects, current theories of international politics were deeply imbued with the American tradition. Moreover, there was a special reason for the ready acceptance of the doctrine in the international sphere. In domestic affairs it is clearly the business of the state to create harmony if no natural harmony exists. In international politics, there is no organised power charged with the task of creating harmony ; and the temptation to assume a natural harmony is therefore particularly strong. But this is no excuse for burking the issue. To make the harmonisation of interests the goal of political action is not the same thing as to postulate that a natural harmony of interests exists ; [1] and it is this latter postulate which has caused so much confusion in international thinking.

The Common Interest in Peace

Politically, the doctrine of the identity of interests has commonly taken the form of an assumption that every nation has an identical interest in peace, and that any nation which desires to disturb the peace is therefore both irrational and immoral. This view bears clear marks of its Anglo-Saxon origin. It was easy after 1918 to convince that part of mankind which lives in English-speaking countries that war profits nobody. The argument did not seem particularly convincing to Germans, who had profited largely from the wars of 1866

---

[1] The confusion between the two was admirably illustrated by an interjection of Mr. Attlee in the House of Commons : " It was precisely the object of the establishment of the League of Nations that the preservation of peace was a common interest of the world " (House of Commons, December 21, 1937 : *Official Report*, col. 1811). Mr. Attlee apparently failed to distinguish between the proposition that a natural community of interests existed and the proposition that the League of Nations had been established to create one.

and 1870, and attributed their more recent sufferings, not to the war of 1914, but to the fact that they had lost it ; or to Italians, who blamed not the war, but the treachery of allies who defrauded them in the peace settlement ; or to Poles or Czecho-Slovaks who, far from deploring the war, owed their national existence to it ; or to Frenchmen, who could not unreservedly regret a war which had restored Alsace-Lorraine to France ; or to people of other nationalities who remembered profitable wars waged by Great Britain and the United States in the past. But these people had fortunately little influence over the formation of current theories of international relations, which emanated almost exclusively from the English-speaking countries. British and American writers continued to assume that the uselessness of war had been irrefutably demonstrated by the experience of 1914–18, and that an intellectual grasp of this fact was all that was necessary to induce the nations to keep the peace in the future ; and they were sincerely puzzled as well as disappointed at the failure of other countries to share this view.

The confusion was increased by the ostentatious readiness of other countries to flatter the Anglo-Saxon world by repeating its slogans. In the fifteen years after the first world war, every Great Power (except, perhaps, Italy) repeatedly did lip-service to the doctrine by declaring peace to be one of the main objects of its policy.[1] But as Lenin observed long ago, peace in itself is a meaningless aim. " Absolutely everybody is in favour of peace in general ", he wrote in 1915, " including Kitchener, Joffre, Hindenburg and Nicholas the Bloody, for everyone of them wishes to end the war." [2] The common interest in peace

[1] " Peace must prevail, must come before all " (Briand, *League of Nations : Ninth Assembly*, p. 83). " The maintenance of peace is the first objective of British foreign policy " (Eden, *League of Nations : Sixteenth Assembly*, p. 106). " Peace is our dearest treasure " (Hitler, in a speech in the German Reichstag on January 30, 1937, reported in *The Times*, February 1, 1937). " The principal aim of the international policy of the Soviet Union is the preservation of peace " (Chicherin in *The Soviet Union and Peace* (1929), p. 249). " The object of Japan, despite propaganda to the contrary, is peace " (Matsuoka, *League of Nations : Special Assembly 1932–33*, iii. p. 73). The paucity of Italian pronouncements in favour of peace was probably explained by the poor reputation of Italian troops as fighters : Mussolini feared that any emphatic expression of preference for peace would be construed as an admission that Italy had no stomach for war.

[2] Lenin, *Collected Works* (Engl. transl.), xviii. p. 264. Compare Spenser Wilkinson's dictum : " It is not peace but preponderance that is in each case the real object. The truth cannot be too often repeated that peace is never the object

masks the fact that some nations desire to maintain the *status quo* without having to fight for it, and others to change the *status quo* without having to fight in order to do so.[1] The statement that it is in the interest of the world as a whole either that the *status quo* should be maintained, or that it should be changed, would be contrary to the facts. The statement that it is in the interest of the world as a whole that the conclusion eventually reached, whether maintenance or change, should be reached by peaceful means, would command general assent, but seems a rather meaningless platitude. The utopian assumption that there is a world interest in peace which is identifiable with the interest of each individual nation helped politicians and political writers everywhere to evade the unpalatable fact of a fundamental divergence of interest between nations desirous of maintaining the *status quo* and nations desirous of changing it.[2] A peculiar combination of platitude and falseness thus became endemic in the pronouncements of statesmen about international affairs. " In this whole Danubian area ", said a Prime Minister of Czecho-Slovakia, " no one really wants conflicts and jealousies. The various countries want to maintain their independence, but otherwise they are ready for any co-operative measures. I am thinking specially of the Little Entente, Hungary and Bulgaria." [3] Literally the words may pass as true. Yet the conflicts and jealousies which nobody wanted were a notorious feature of Danubian politics after 1919, and the co-operation for which all were ready was unobtainable. The fact of divergent interests was disguised and falsified by the platitude of a general desire to avoid conflict.

of policy : you cannot define peace except by reference to war, which is a means and never an end " (*Government and the War*, p. 121).

[1] " When a saint complains that people do not know the things belonging to their peace, what he really means is that they do not sufficiently care about the things belonging to his peace " (*The Note-Books of Samuel Butler*, ed. Festing-Jones, pp. 211-12). This would seem to be true of those latter-day saints, the satisfied Powers.

[2] It is sometimes maintained not merely that all nations have an equal interest in preferring peace to war (which is, in a sense, true), but that war can never in any circumstances bring to the victor advantages comparable with its cost. The latter view does not appear to be true of the past, though it is possible to argue (as does Bertrand Russell, *Which Way Peace ?*) that it is true of modern warfare. If accepted, this view leads, of course, to absolute pacifism ; for there is no reason to suppose that it is any truer of " defensive " than of " offensive " war (assuming the distinction between them to be valid).

[3] *Daily Telegraph*, August 26, 1938.

International Economic Harmony

In economic relations, the assumption of a general harmony of interests was made with even greater confidence ; for here we have a direct reflexion of the cardinal doctrine of *laissez-faire* economics, and it is here that we can see most clearly the dilemma which results from the doctrine. When the nineteenth-century liberal spoke of the greatest good of the greatest number, he tacitly assumed that the good of the minority might have to be sacrificed to it. This principle applied equally to international economic relations. If Russia or Italy, for example, were not strong enough to build up industries without the protection of tariffs, then — the *laissez-faire* liberal would have argued — they should be content to import British and German manufactures and supply wheat and oranges to the British and German markets. If anyone had thereupon objected that this policy would condemn Russia and Italy to remain second-rate Powers economically and militarily dependent on their neighbours, the *laissez-faire* liberal would have had to answer that this was the will of Providence and that this was what the general harmony of interests demanded. The modern utopian internationalist enjoys none of the advantages, and has none of the toughness, of the nineteenth-century liberal. The material success of the weaker Powers in building up protected industries, as well as the new spirit of internationalism, preclude him from arguing that the harmony of interests depends on the sacrifice of economically unfit nations. Yet the abandonment of this premiss destroys the whole basis of the doctrine which he has inherited ; and he is driven to the belief that the common good can be achieved without any sacrifice of the good of any individual member of the community. Every international conflict is therefore unnecessary and illusory. It is only necessary to discover the common good which is at the same time the highest good of all the disputants ; and only the folly of statesmen stands in the way of its discovery. The utopian, secure in his understanding of this common good, arrogates to himself the monopoly of wisdom. The statesmen of the world one and all stand convicted of incredible blindness to the interest of those whom they are supposed to represent. Such was the picture of the international scene presented, in all seriousness, by British and American writers, including not a few economists.

It is for this reason that we find in the modern period an extraordinary divergence between the theories of economic experts and the practice of those responsible for the economic policies of their respective countries. Analysis will shew that this divergence springs from a simple fact. The economic expert, dominated in the main by *laissez-faire* doctrine, considers the hypothetical economic interest of the world as a whole, and is content to assume that this is identical with the interest of each individual country. The politician pursues the concrete interest of his country, and assumes (if he makes any assumption at all) that the interest of the world as a whole is identical with it. Nearly every pronouncement of every international economic conference held between the two world wars was vitiated by this assumption that there was some " solution " or " plan " which, by a judicious balancing of interests, would be equally favourable to all and prejudicial to none.

Any strictly nationalistic policy [declared the League Conference of economic experts in 1927] is harmful not only to the nation which practises it but also to the others, and therefore defeats its own end, and if it be desired that the new state of mind revealed by the Conference should lead rapidly to practical results, any programme of execution must include, as an essential factor, the principle of *parallel* or *concerted* action by the different nations. Every country will then know that the concessions it is asked to make will be balanced by corresponding sacrifices on the part of the other countries. It will be able to accept the proposed measures, not merely in view of its own individual position, *but also because it is interested in the success of the general plan laid down by the Conference.*[1]

The sequel of the Conference was the complete neglect of all the recommendations unanimously made by it ; and if we are not content to accept the facile explanation that the leading statesmen of the world were either criminal or mad, we may begin to suspect the validity of its initial assumption. It seems altogether rash to suppose that economic nationalism is necessarily detrimental to states which practise it. In the nineteenth century, Germany and the United States, by pursuing a

[1] *League of Nations*: C.E.I. 44, p. 21 (italics in original).

" strictly nationalistic policy ", had placed themselves in a position to challenge Great Britain's virtual monopoly of world trade. No conference of economic experts, meeting in 1880, could have evolved a " general plan " for "parallel or concerted action " which would have allayed the economic rivalries of the time in a manner equally advantageous to Great Britain, Germany and the United States. It was not less presumptuous to suppose that a conference meeting in 1927 could allay the economic rivalries of the later period by a "plan" beneficial to the interests of everyone. Even the economic crisis of 1930–33 failed to bring home to the economists the true nature of the problem which they had to face. The experts who prepared the " Draft Annotated Agenda " for the World Economic Conference of 1933 condemned the " world-wide adoption of ideals of national self-sufficiency which cut unmistakably athwart the lines of economic development ".[1] They did not apparently pause to reflect that those so-called " lines of economic development ", which might be beneficial to some countries and even to the world as a whole, would inevitably be detrimental to other countries, which were using weapons of economic nationalism in self-defence. The Van Zeeland report of January 1938 began by asking, and answering in the affirmative, the question whether " the methods which, taken as a whole, form the system of international trade " are "fundamentally preferable " to " autarkic tendencies ". Yet every Power at some period of its history, and as a rule for prolonged periods, has resorted to "autarkic tendencies ". It is difficult to believe that there is any absolute sense in which "autarkic tendencies " are always detrimental to those who pursue them. Even if they could be justified only as the lesser of two evils, the initial premise of the Van Zeeland report was invalidated. But there was worse to come. "We must . . . make our dispositions", continued M. Van Zeeland, "in such a way that the new system shall offer to all participators advantages greater than those of the position in which they now find themselves." [2] This is economic utopianism in its most purblind form. The report, like the reports of 1927 and 1933, assumed the existence of a fundamental principle of economic

[1] *League of Nations* : C.48, M.18, 1933, ii. p. 6.
[2] *Report . . . on the Possibility of Obtaining a General Reduction of the Obstacles to International Trade*, Cmd. 5648.

policy whose application would be equally beneficial to all states and detrimental to none ; and for this reason it remained, like its predecessors, a dead letter.

Economic theory, as opposed to economic practice, was so powerfully dominated in the years between the two world wars by the supposed harmony of interests that it is difficult to find, in the innumerable international discussions of the period, any clear exposition of the real problem which baffled the statesmen of the world. Perhaps the frankest statement was one made by the Yugoslav Foreign Minister at the session of the Commission for European Union in January 1931. Arthur Henderson, on behalf of Great Britain, following the Netherland delegate Dr. Colijn, had pleaded for an all-round tariff reduction " which must, by its nature, bring benefit to each and all by allowing that expansion of production and international exchange of wealth by which the common prosperity of all can be increased ".[1] Marinkovitch, who spoke next, concluded from the failure to carry out the recommendations of the 1927 Conference, that " there were extremely important reasons why the governments could not apply " those resolutions. He went on :

> The fact is that apart from economic considerations there are also political and social considerations. The old "things-will-right-themselves " school of economists argued that if nothing were done and events were allowed to follow their natural course from an economic point of view, economic equilibrium would come about of its own accord. That is probably true (I do not propose to discuss the point). But how would that equilibrium come about ? At the expense of the weakest. Now, as you are aware, for more than seventy years there has been a powerful and growing reaction against this theory of economics. All the socialist parties of Europe and the world are merely the expression of the opposition to this way of looking at economic problems.
>
> We were told that we ought to lower customs barriers and even abolish them. As far as the agricultural states of Europe are concerned, if they could keep the promises they made in 1927 — admitting that the statements of 1927 did contain promises — and could carry that policy right through, we might perhaps find ourselves able to hold our

[1] *League of Nations* : C.144, M.45, 1931, vii. p. 30.

own against overseas competition in the matter of agricultural products. But at the same time we should have to
create in Poland, Roumania and Yugoslavia the same conditions as exist in Canada and the Argentine, where vast
territories are inhabited by a scanty population and where
machinery and other devices are employed. . . . We could
not sacrifice our people by shooting them, but they would
nevertheless be killed off by famine — which would come to
the same thing.

I am sure that the key to which M. Colijn has referred
does not exist. Economic and social life is too complicated
to allow of a solution by any one formula ; it calls for complicated solutions. We shall have to take into account the
many varieties of geographical, political, social and other
conditions which exist.[1]

Marinkovitch went on to dispose of the theory of the " long-
run " harmony of interests :

Last year, when I was in the Yugoslav mountains, I heard
that the inhabitants of a small mountain village, having no
maize or wheat on which to live, were simply cutting down a
wood which belonged to them . . . and were living on what
they earned by selling the wood. . . . I went to the village,
collected together some of the leading inhabitants and endeavoured to reason with them, just like the great industrial states reason with us. I said to them : " You possess
plenty of common sense. You see that your forest is becoming
smaller and smaller. What will you do when you cut down
the last tree ? " They replied to me : " Your Excellency,
that is a point which worries us : but on the other hand,
what should we do now if we stopped cutting down our
trees ? "

I can assure you that the agricultural countries are in
exactly the same situation. You threaten them with future
disasters ; but they are already in the throes of disaster.[2]

One further example of unwonted frankness may be quoted.
Speaking in September 1937 over one of the United States
broadcasting systems, the President of the Colombian Republic
said :

In no field of human activity are the benefits of the crisis
as clear as in the relationships between nations and especially

[1] *League of Nations*: C.144, M.45, 1931, vii. p. 31.   [2] *Ibid.* p. 32.

of the American nations. If it is true that the economic relations have become rigorous and at times harsh, it is also true that they have fortunately become more democratic.

The crisis freed many countries which had up to then been subordinated to the double mental and financial imperialism of the nations which controlled international markets and policies. Many nations learned to trust less international cordiality and to seek an autonomous life, full of initial obstacles but which nevertheless created strong interests within a short time. . . .

When the arbitrary systems that prevail to-day begin to be relaxed, there will be a weaker international trade, but there will also be a larger number of nations economically strong.

Economic co-operation to-day is a very different and more noble thing than the old co-operation which was based on the convenience of industrial countries and of bankers who tutored the world. The certainty acquired by many small nations that they can subsist and prosper without subordinating their conduct and their activities to foreign interests has begun to introduce a greater frankness and equality in the relations between modern nations. . . .

It is true that the crisis has shipwrecked many high and noble principles of our civilisation ; but it is also true that in this return to a kind of primitive struggle for existence, peoples are being freed of many fictions and of much hypocrisy which they had accepted in the belief that with them they were insuring their well-being. . . .

The foundation of international economic freedom lies in the recognition that when strong nations place themselves on the defensive, they act just like the weak ones do, and that all of them have an equal right to defend themselves with their own resources.[1]

The claims made on behalf of the Colombian Republic were perhaps exaggerated. But both the Yugoslav and the Colombian statements were powerful challenges to the doctrine of the harmony of interests. It is fallacy to suppose that, because Great Britain and the United States have an interest in the removal of trade barriers, this is also an interest of Yugoslavia and Colombia. International trade may be weaker. The

[1] Address broadcast by the Columbia Broadcasting System, U.S.A., on September 19, 1937, and published in *Talks*, October 1937.

economic interests of Europe or of the world at large may
suffer. But Yugoslavia and Colombia will be better off than
they would have been under a régime of European or world
prosperity which reduced them to the position of satellites. Dr.
Schacht spoke a little later of those " fanatical adherents of
the most-favoured-nation policy abroad, who from the abun-
dance of their wealth cannot realise that a poor nation has
nevertheless the courage to live by its own laws instead of
suffering under the prescriptions of the well-to-do ".[1] *Laissez-
faire*, in international relations as in those between capital and
labour, is the paradise of the economically strong. State control, whether in the form of protective legislation or of protective
tariffs, is the weapon of self-defence invoked by the economically
weak. The clash of interests is real and inevitable ; and the
whole nature of the problem is distorted by an attempt to
disguise it.

The Harmony Broken

We must therefore reject as inadequate and misleading the
attempt to base international morality on an alleged harmony
of interests which identifies the interest of the whole community
of nations with the interest of each individual member of it.
In the nineteenth century, this attempt met with widespread
success, thanks to the continuously expanding economy in
which it was made. The period was one of progressive prosperity, punctuated only by minor set-backs. The international economic structure bore considerable resemblance to
the domestic economic structure of the United States. Pressure
could at once be relieved by expansion to hitherto unoccupied
and unexploited territories ; and there was a plentiful supply
of cheap labour, and of backward countries, which had not
yet reached the level of political consciousness. Enterprising
individuals could solve the economic problem by migration,
enterprising nations by colonisation. Expanding markets produced an expanding population, and population in turn reacted
on markets. Those who were left behind in the race could
plausibly be regarded as the unfit. A harmony of interests
among the fit, based on individual enterprise and free competition, was sufficiently near to reality to form a sound basis

[1] Address to the Economic Council of the German Academy, November 29,
1938.

for the current theory. With some difficulty the illusion was kept alive till 1914. Even British prosperity, though its foundations were menaced by German and American competition, continued to expand. The year 1913 was a record year for British trade.

The transition from the apparent harmony to the transparent clash of interests may be placed about the turn of the century. Appropriately enough, it found its first expression in colonial policies. In the British mind, it was primarily associated with events in South Africa. Mr. Churchill dates the beginning of " these violent times " from the Jameson Raid.[1] In North Africa and the Far East, there was a hasty scramble by the European Powers to secure the few eligible sites which were still vacant. Emigration of individuals from Europe, the point of principal tension, to America assumed unparalleled dimensions. In Europe itself, anti-Semitism — the recurrent symptom of economic stress — reappeared after a long interval in Russia, Germany and France.[2] In Great Britain, agitation against unrestricted alien immigration began in the 1890's ; and the first act controlling immigration was passed in 1905.

The first world war, which proceeded from this growing tension, aggravated it tenfold by intensifying its fundamental causes. In belligerent and neutral countries in Europe, Asia and America, industrial and agricultural production were everywhere artificially stimulated. After the war every country struggled to maintain its expanded production ; and an enhanced and inflamed national consciousness was invoked to justify the struggle. One reason for the unprecedented vindictiveness of the peace treaties, and in particular of their economic clauses, was that practical men no longer believed — as they had done fifty or a hundred years earlier — in an underlying harmony of interests between victors and defeated. The object was now to eliminate a competitor, a revival of whose prosperity might menace your own. In Europe, the struggle was intensified by the creation of new states and new economic frontiers. In Asia, India and China built up large-scale manufactures to make themselves independent of imports

---

[1] Winston Churchill, *World Crisis*, p. 26.

[2] The same conditions encouraged the growth of Zionism ; for Zionism, as the Palestine Royal Commission of 1937 remarked, " on its negative side is a creed of escape " Cmd. 5479, p. 13).

from Europe. Japan became an exporter of textiles and other cheap goods which undercut European manufactures on the world market. Most important of all, there were no more open spaces anywhere awaiting cheap and profitable development and exploitation. The ample avenues of migration which had relieved the economic pressures of the pre-war period were closed ; and in place of the natural flow of migration came the problem of forcibly evicted refugees.[1] The complex phenomenon known as economic nationalism swept over the world. The fundamental character of this clash of interests became obvious to all except those confirmed utopians who dominated economic thought in the English-speaking countries. The hollowness of the glib nineteenth-century platitude that nobody can benefit from what harms another was revealed. The basic presupposition of utopianism had broken down.

What confronts us in international politics to-day is, therefore, nothing less than the complete bankruptcy of the conception of morality which has dominated political and economic thought for a century and a half. Internationally, it is no longer possible to deduce virtue from right reasoning, because it is no longer seriously possible to believe that every state, by pursuing the greatest good of the whole world, is pursuing the greatest good of its own citizens, and *vice versa*. The synthesis of morality and reason, at any rate in the crude form in which it was achieved by nineteenth-century liberalism, is untenable. The inner meaning of the modern international crisis is the collapse of the whole structure of utopianism based on the concept of the harmony of interests. The present generation will have to rebuild from the foundations. But before we can do this, before we can ascertain what can be salved from the ruins, we must examine the flaws in the structure which led to its collapse ; and we can best do this by analysing the realist critique of the utopian assumptions.

[1] " The existence of refugees is a symptom of the disappearance of economic and political liberalism. Refugees are the by-product of an economic isolationism which has practically stopped free migration " (J. Hope Simpson, *Refugees: Preliminary Report of a Survey*, p. 193).

CHAPTER 5

THE REALIST CRITIQUE

The Foundations of Realism

FOR reasons explained in a previous chapter, realism enters the field far behind utopianism and by way of reaction from it. The thesis that "justice is the right of the stronger" was, indeed, familiar in the Hellenic world. But it never represented anything more than the protest of an uninfluential minority, puzzled by the divergence between political theory and political practice. Under the supremacy of the Roman Empire, and later of the Catholic Church, the problem could hardly arise ; for the political good, first of the empire, then of the church, could be regarded as identical with moral good. It was only with the break-up of the mediaeval system that the divergence between political theory and political practice became acute and challenging. Machiavelli is the first important political realist.

Machiavelli's starting-point is a revolt against the utopianism of current political thought :

> It being my intention to write a thing which shall be useful to him who apprehends it, it appears to me more appropriate to follow up the real truth of a matter than the imagination of it ; for many have pictured republics and principalities which in fact have never been seen and known, because how one lives is so far distant from how one ought to live that he who neglects what is done for what ought to be done sooner effects his ruin than his preservation.

The three essential tenets implicit in Machiavelli's doctrine are the foundation-stones of the realist philosophy. In the first place, history is a sequence of cause and effect, whose course can be analysed and understood by intellectual effort, but not (as the utopians believe) directed by "imagination". Secondly, theory does not (as the utopians assume) create practice, but practice theory. In Machiavelli's words, " good counsels, whencesoever they come, are born of the wisdom of the prince, and not the wisdom of the prince from good counsels ". Thirdly, politics are not (as the utopians pretend) a function of ethics, but ethics of politics. Men " are kept honest by constraint ". Machiavelli recognised the importance of morality, but thought that there could be no effective morality where there was no effective authority. Morality is the product of power.[1]

The extraordinary vigour and vitality of Machiavelli's challenge to orthodoxy may be attested by the fact that, more than four centuries after he wrote, the most conclusive way of discrediting a political opponent is still to describe him as a disciple of Machiavelli.[2] Bacon was one of the first to praise him for " saying openly and without hypocrisy what men are in the habit of doing, not what they ought to do ".[3] Henceforth no political thinker could ignore him. In France Bodin, in England Hobbes, in the Netherlands Spinoza, professed to find a compromise between the new doctrine and the conception of a " law of nature " constituting a supreme ethical standard. But all three were in substance realists ; and the age of Newton for the first time conceived the possibility of a physical science of politics.[4] The work of Bodin and Hobbes, writes Professor Laski, was " to separate ethics from politics, and to complete by theoretical means the division which Machiavelli had effected on practical grounds ".[5] " Before the names of Just and Unjust can have place ", said Hobbes, " there must be some coercive power." [6] Spinoza believed that practical statesmen

---

[1] Machiavelli, *The Prince*, chs. 15 and 23 (Engl. transl., Everyman's Library, pp. 121, 193).

[2] Two curious recent illustrations may be cited. In the chapter of the *Survey of International Affairs* dealing with the Nazi revolution, Professor Toynbee declares that National Socialism is the "fulfilment of ideals . . . formulated . . . by Machiavelli " ; and he reiterates this view in two further passages of considerable length in the same chapter (*Survey of International Affairs, 1934*, pp. 111, 117-19, 126-8). In the trial of Zinoviev, Kamenev and others in Moscow in August 1936, the Public Prosecutor, Vyshinsky, quoted a passage from Kamenev's writings in which Machiavelli had been praised as "a master of political aphorism and a brilliant dialectician ", and accused Kamenev of having " adopted the rules of Machiavelli " and " developed them to the utmost point of unscrupulousness and immorality " (*The Case of the Trotskyite-Zinovievite Centre*, pp. 138-9).

[3] Bacon, *On the Advancement of Learning*, vii. ch. 2.

[4] Hobbes's scheme, " there was in theory no place for any new force or principle beyond the laws of motion found at the beginning ; there were merely complex cases of mechanical causation " (Sabine, *History of Political Thought*, p. 458).

[5] Introduction to *A Defence of Liberty against Tyrants (Vindiciae contra Tyrannos)*, ed. Laski, p. 45.

[6] Hobbes, *Leviathan*, ch. xv.

had contributed more to the understanding of politics than men of theory " and, above all, theologians " ; for " they have put themselves to the school of experience, and have therefore taught nothing which does not bear upon our practical needs ".[1] In anticipation of Hegel, Spinoza declares that " every man does what he does according to the laws of his nature and to the highest right of nature ".[2] The way is thus opened for determinism ; and ethics become, in the last analysis, the study of reality.

Modern realism differs, however, in one important respect from that of the sixteenth and seventeenth centuries. Both utopianism and realism accepted and incorporated in their philosophies the eighteenth-century belief in progress, with the curious and somewhat paradoxical result that realism became in appearance more " progressive " than utopianism. Utopianism grafted its belief in progress on to its belief in an absolute ethical standard, which remained *ex hypothesi* static. Realism, having no such sheet-anchor, became more and more dynamic and relativist. Progress became part of the inner essence of the historical process ; and mankind was moving forward towards a goal which was left undefined, or was differently defined by different philosophers. The " historical school " of realists had its home in Germany, and its development is traced through the great names of Hegel and Marx. But no country in Western Europe, and no branch of thought, was immune from its influence in the middle and later years of the nineteenth century ; and this development, while it has freed realism from the pessimistic colouring imparted to it by thinkers like Machiavelli and Hobbes, has thrown its determinist character into stronger relief.

The idea of causation in history is as old as the writing of history itself. But so long as the belief prevailed that human affairs were subject to the continuous supervision and occasional intervention of a Divine Providence, no philosophy of history based on a regular relationship of cause and effect was likely to be evolved. The substitution of reason for Divine Providence enabled Hegel to produce, for the first time, a philosophy based on the conception of a rational historical process. Hegel, while assuming a regular and orderly process, was content to find its directing force in a metaphysical abstraction — the *Zeitgeist*.

[1] Spinoza, *Tractatus Politicus*, i pp. 2-3.　　[2] *Ibid.* Introduction.

But once the historical conception of reality had established itself, it was a short step to substitute for the abstract *Zeitgeist* some concrete material force. The economic interpretation of history was not invented, but developed and popularised, by Marx. About the same time Buckle propounded a geographical interpretation of history which convinced him that human affairs were " permeated by one glorious principle of universal and undeviating regularity " ; [1] and this has been revived in the form of the science of *Geopolitik*, whose inventor describes geography as " a political categorical imperative ".[2] Spengler believed that events were determined by quasi-biological laws governing the growth and decline of civilisations. More eclectic thinkers interpret history as the product of a variety of material factors, and the policy of a group or nation as a reflexion of all the material factors which make up the group or national interest. " Foreign policies ", said Mr. Hughes during his tenure of office as American Secretary of State, " are not built upon abstractions. They are the result of national interest arising from some immediate exigency or standing out vividly in historical perspective." [3] Any such interpretation of reality, whether in terms of a *Zeitgeist*, or of economics or geography, or of " historical perspective ", is in its last analysis deterministic. Marx (though, having a programme of action, he could not be a rigid and consistent determinist) believed in " tendencies which work out with an iron necessity towards an inevitable goal ".[4] " Politics ", wrote Lenin, " have their own objective logic independent of the prescriptions of this or that individual or party." [5] In January 1918, he described his belief in the coming socialist revolutions in Europe as " a scientific prediction ".[6]

On the " scientific " hypothesis of the realists, reality is thus identified with the whole course of historical evolution, whose laws it is the business of the philosopher to investigate and

[1] The concluding words of Buckle's *History of Civilisation*.
[2] Kjellen, *Der Staat als Lebensform*, p. 81. Compare the opening words of Crowe's famous memorandum on British foreign policy : " The general character of England's foreign policy is determined by the immutable conditions of her geographical situation " (*British Documents on the Origin of the War*, ed. Gooch and Temperley, iii. p. 397).
[3] *International Conciliation*, No. 194, January 1924, p. 3.
[4] Marx, *Capital*, Preface to 1st ed. (Engl. transl., Everyman's Library, p. 863).
[5] Lenin, *Works* (2nd Russian ed.), x. p. 207.
[6] *Ibid.* xxii. p. 194.

reveal. There can be no reality outside the historical process. " To conceive of history as evolution and progress ", writes Croce, " implies accepting it as necessary in all its parts, and therefore denying validity to judgments on it." [1] Condemnation of the past on ethical grounds has no meaning ; for in Hegel's words, " philosophy transfigures the real which appears unjust into the rational ".[2] What was, is right. History cannot be judged except by historical standards. It is significant that our historical judgments, except those relating to a past which we can ourselves remember as the present, always appear to start from the presupposition that things could not have turned out otherwise than they did. It is recorded that Venizelos, on reading in Fisher's *History of Europe* that the Greek invasion of Asia Minor in 1919 was a mistake, smiled ironically and said : " Every enterprise that does not succeed is a mistake ".[3] If Wat Tyler's rebellion had succeeded, he would be an English national hero. If the American War of Independence had ended in disaster, the Founding Fathers of the United States would be briefly recorded in history as a gang of turbulent and unscrupulous fanatics. Nothing succeeds like success. " World history ", in the famous phrase which Hegel borrowed from Schiller, " is the world court ". The popular paraphrase " Might is Right " is misleading only if we attach too restricted a meaning to the word "Might". History creates rights, and therefore right. The doctrine of the survival of the fittest proves that the survivor was, in fact, the fittest to survive. Marx does not seem to have maintained that the victory of the proletariat was just in any other sense than that it was historically inevitable. Lukacs was a consistent, though perhaps indiscreet, Marxist when he based the " right " of the proletariat on its " historical mission ".[4] Hitler believed in the historical mission of the German people.

The Relativity of Thought

The outstanding achievement of modern realism, however, has been to reveal, not merely the determinist aspects of the historical process, but the relative and pragmatic character of

[1] Croce, *Storia della storiografia italiana*, i. p. 26.
[2] Hegel, *Philosophie der Weltgeschichte* (Lasson's ed.), p. 55.
[3] *Conciliation Internationale*, No. 5-6, 1937, p. 520.
[4] Lukacs, *Geschichte und Klassenbewusstsein*, p. 215.

thought itself. In the last fifty years, thanks mainly though not wholly to the influence of Marx, the principles of the historical school have been applied to the analysis of thought; and the foundations of a new science have been laid, principally by German thinkers, under the name of the " sociology of knowledge ". The realist has thus been enabled to demonstrate that the intellectual theories and ethical standards of utopianism, far from being the expression of absolute and *a priori* principles, are historically conditioned, being both products of circumstances and interests and weapons framed for the furtherance of interests. " Ethical notions ", as Mr. Bertrand Russell has remarked, " are very seldom a cause, but almost always an effect, a means of claiming universal legislative authority for our own preferences, not, as we fondly imagine, the actual ground of those preferences." [1] This is by far the most formidable attack which utopianism has to face; for here the very foundations of its belief are undermined by the realist critique.

In a general way, the relativity of thought has long been recognised. As early as the seventeenth century Bishop Burnet expounded the relativist view as cogently, if not as pungently, as Marx :

As to the late Civil Wars, 'tis pretty well known what notions of government went current in those days. When monarchy was to be subverted we knew what was necessary to justify the fact; and then, because it was convenient for the purpose, it was undoubtedly true in the nature of things that government had its original from the people, and the prince was only their trustee. . . . But afterwards, when monarchy took its place again . . . another notion of government came into fashion. Then government had its original entirely from God, and the prince was accountable to none but Him. . . . And now, upon another turn of things, when people have a liberty to speak out, a new set of notions is advanced; now passive obedience is all a mistake, and instead of being a duty to suffer oppression, 'tis a glorious act to resist it : and instead of leaving injuries to be redressed by God, we have a natural right to relieve ourselves.[2]

[1] *Proceedings of the Aristotelian Society*, 1915–16, p. 302
[2] Burnet, *Essay upon Government*, p. 10.

In modern times, the recognition of this phenomenon has become fairly general. " Belief, and to speak fairly, honest belief ", wrote Dicey of the divisions of opinion in the nineteenth century about slavery, " was to a great extent the result not of argument, not even of direct self-interest, but of circumstances. . . . Circumstances are the creators of most men's opinions." [1] Marx narrowed down this somewhat vague conception, declaring that all thought was conditioned by the economic interest and social status of the thinker. This view was perhaps unduly restrictive. In particular Marx, who denied the existence of " national " interests, underestimated the potency of nationalism as a force conditioning the thought of the individual. But the peculiar concentration which he applied to the principle served to popularise it and drive it home. The relativity of thought to the interests and circumstances of the thinker has been far more extensively recognised and understood since Marx wrote.

The principle has an extremely wide field of application. It has become a commonplace to say that theories do not mould the course of events, but are invented to explain them. " Empire precedes imperialism." [2] Eighteenth-century England " put into practice the policy of *laissez-faire* before it found a justification, or even an apparent justification, in the new doctrine " ; [3] and " the virtual break-up of *laissez-faire* as a body of doctrine . . . has followed, and not preceded, the decline of *laissez-faire* in the real world ".[4] The theory of " socialism in a single country " promulgated in Soviet Russia in 1924 was manifestly a product of the failure of Soviet régimes to establish themselves in other countries.

But the development of abstract theory is often influenced by events which have no essential connexion with it at all.

In the story of political thought [writes a modern social thinker] events have been no less potent than arguments. The failure and success of institutions, the victories and defeats of countries identified with certain principles have repeatedly brought new strength and resolution to the adherents or opponents of these principles as the case might

---

[1] Dicey, *Law and Opinion* (1905 ed.), p. 27.
[2] J. A. Hobson, *Free Thought in the Social Sciences*, p. 190.
[3] Halévy, *The Growth of Philosophic Radicalism* (Engl. transl.), p. 104.
[4] M. Dobb, *Political Economy and Capitalism*, p. 188.

be in all lands. . . . Philosophy as it exists on earth is the word of philosophers who, authority tells us, suffer as much from toothache as other mortals, and are, like others, open to the impression of near and striking events and to the seductions of intellectual fashion.[1]

Germany's dramatic rise to power in the sixties and seventies of last century was impressive enough to make the leading British philosophers of the next generation — Caird, T. H. Green, Bosanquet, McTaggart — ardent Hegelians. Thereafter, the Kaiser's telegram to Kruger and the German naval programme spread the conviction among British thinkers that Hegel was a less good philosopher than had been supposed ; and since 1914 no British philosopher of repute has ventured to sail under the Hegelian flag. After 1870, Stubbs and Freeman put early English history on a sound Teutonic basis, while even in France Fustel de Coulanges had an uphill struggle to defend the Latin origins of French civilisation. During the past thirty years, English historians have been furtively engaged in making the Teutonic origins of England as inconspicuous as possible.

Nor is it only professional thinkers who are subject to such influences. Popular opinion is not less markedly dominated by them. The frivolity and immorality of French life was an established dogma in nineteenth-century Britain, which still remembered Napoleon. "When I was young", writes Mr. Bertrand Russell, "the French ate frogs and were called 'froggies', but they apparently abandoned this practice when we concluded our *entente* with them in 1904 — at any rate, I have never heard it mentioned since that date." [2] Some years later, "the gallant little Jap" of 1905 underwent a converse metamorphosis into "the Prussian of the East". In the nineteenth century, it was a commonplace of British opinion that Germans were efficient and enlightened, and Russians backward and barbarous. About 1910, it was ascertained that Germans (who turned out to be mostly Prussians) were coarse, brutal and narrow-minded, and that Russians had a Slav soul. The vogue of Russian literature in Great Britain, which set in about the same time, was a direct outcome of the political

[1] L. T. Hobhouse, *The Unity of Western Civilisation*, ed. F. S. Marvin (3rd ed.), pp. 177-8.
[2] Bertrand Russell, *Which Way Peace ?* p. 158.

rapprochement with Russia. The vogue of Marxism in Great Britain and France, which began on a modest scale after the success of the Bolshevik revolution in Russia, rapidly gathered momentum, particularly among intellectuals, after 1934, when it was discovered that Soviet Russia was a potential military ally against Germany. It is symptomatic that most people, when challenged, will indignantly deny that they form their opinions in this way ; for as Acton long ago observed, " few discoveries are more irritating than those which expose the pedigree of ideas ".[1] The conditioning of thought is necessarily a subconscious process

The Adjustment of Thought to Purpose

Thought is not merely relative to the circumstances and interests of the thinker : it is also pragmatic in the sense that it is directed to the fulfilment of his purposes. For the realist, as a witty writer has put it, truth is " no more than the perception of discordant experience pragmatically adjusted for a particular purpose and for the time being ".[2] The purposeful character of thought has been discussed in a previous chapter ; and a few examples will suffice here to illustrate the importance of this phenomenon in international politics.

Theories designed to discredit an enemy or potential enemy are one of the commonest forms of purposeful thinking. To depict one's enemies or one's prospective victims as inferior beings in the sight of God has been a familiar technique at any rate since the days of the Old Testament. Racial theories, ancient and modern, belong to this category ; for the rule of one people or class over another is always justified by a belief in the mental and moral inferiority of the ruled. In such theories, sexual abnormality and sexual offences are commonly imputed to the discredited race or group. Sexual depravity is imputed by the white American to the negro ; by the white South African to the Kaffir ; by the Anglo-Indian to the Hindu ; and by the Nazi German to the Jew. The most popular and most absurd of the charges levelled against the Bolsheviks in the early days of the Russian revolution was that they advocated sexual promiscuity. Atrocity stories,

---

[1] Acton, *History of Freedom*, p. 62.
[2] Carl Becker, *Yale Review*, xxvii, p. 461.

among which offences of a sexual character predominate, are
the familiar product of war. On the eve of their invasion
of Abyssinia, the Italians issued an official Green Book of
Abyssinian atrocities. " The Italian Government ", as the
Abyssinian delegate at Geneva correctly observed, " having
resolved to conquer and destroy Ethiopia, begins by giving
Ethiopia a bad name."[1]

But the phenomenon also appears in less crude forms which
sometimes enable it to escape detection. The point was well
made by Crowe in a Foreign Office minute of March 1908 :

> The German (formerly Prussian) Government has always
> been most remarkable for the pains it takes to create a feeling
> of intense and holy hatred against a country with which it
> contemplates the possibility of war. It is undoubtedly in
> this way that the frantic hatred of England as a monster of
> personified selfishness and greed and absolute want of con-
> science, which now animates Germany, has been nursed and
> fed [2]

The diagnosis is accurate and penetrating. But it is strange
that so acute a mind as Crowe's should not have perceived that
he himself was at this time performing, for the limited audience
of statesmen and officials to which he had access, precisely the
same operation of which he accused the German Government ;
for a perusal of his memoranda and minutes of the period
reveals an able, but transparent, attempt to " create a feeling
of intense and holy hatred " against his own country's future
enemy — a curious instance of our promptness to detect the
conditioned or purposeful character of other people's thought,
while assuming that our own is wholly objective.

The converse of this propagation of theories designed to
throw moral discredit on an enemy is the propagation of theories
reflecting moral credit on oneself and one's own policies. Bismarck records the remark made to him by Walewski, the
French Foreign Minister, in 1857, that it was the business of
a diplomat to cloak the interests of his country in the language
of universal justice. More recently, Mr. Churchill told the
House of Commons that " there must be a moral basis for

---

[1] *League of Nations : Official Journal*, November 1935, p. 1140.
[2] *British Documents on the Origins of the War*, ed. Gooch and Temperley,
vi. p. 131.

British rearmament and foreign policy ".[1] It is rare, however, for modern statesmen to express themselves with this frankness ; and in contemporary British and American politics, the most powerful influence has been wielded by those more utopian statesmen who are sincerely convinced that policy is deduced from ethical principles, not ethical principles from policy. The realist is nevertheless obliged to uncover the hollowness of this conviction. " The right ", said Woodrow Wilson to the United States Congress in 1917, " is more precious than peace."[2] " Peace comes before all," said Briand ten years later to the League of Nations Assembly, " peace comes even before ˙justice."[3] Considered as ethical principles, both these contradictory pronouncements are tenable and could muster respectable support. Are we therefore to believe that we are dealing with a clash of ethical standards, and that if Wilson's and Briand's policies differed it was because they deduced them from opposite principles ? No serious student of politics will entertain this belief. The most cursory examination shews that the principles were deduced from the policies, not the policies from the principles. In 1917, Wilson had decided on the policy of war with Germany, and he proceeded to clothe that policy in the appropriate garment of righteousness. In 1928 Briand was fearful of attempts made in the name of justice to disturb a peace settlement favourable to France ; and he had no more difficulty than Wilson in finding the moral phraseology which fitted his policy. It would be irrelevant to discuss this supposed difference of principles on ethical grounds. The principles merely reflected different national policies framed to meet different conditions.

The double process of morally discrediting the policy of a potential enemy and morally justifying one's own may be abundantly illustrated from the discussions of disarmament between the two wars. The experience of the Anglo-Saxon Powers, whose naval predominance had been threatened by the submarine, provided an ample opportunity of denouncing the immorality of this new weapon. " Civilisation demands ", wrote the naval adviser to the American Delegation at the

---

[1] House of Commons, March 14, 1938 : *Official Report*, cols. 95-99.
[2] *The Public Papers of Woodrow Wilson : War and Peace*, ed. R. S. Baker, i. p. 16.
[3] *League of Nations : Ninth Assembly*, p. 83.

Peace Conference, " that naval warfare be placed on a higher plane " by the abolition of the submarine.[1] Unfortunately the submarine was regarded as a convenient weapon by the weaker French, Italian and Japanese navies ; and this particular demand of civilisation could not therefore be complied with. A distinction of a more sweeping character was established by Lord Cecil in a speech to the General Council of the League of Nations Union in 1922 :

> The general peace of the world will not be materially secured merely by naval disarmament. . . . If all the maritime Powers were to disarm, or drastically limit their armaments, I am not at all sure that would not increase the danger of war rather than decrease it, because the naval arm is mainly defensive ; the offensive must be to a large extent the military weapon.[2]

The inspiration of regarding one's own vital armaments as defensive and beneficent and those of other nations as offensive and wicked proved particularly fruitful. Exactly ten years later, three commissions of the Disarmament Conference spent many weeks in a vain endeavour to classify armaments as " offensive " and " defensive ". Delegates of all nations shewed extraordinary ingenuity in devising arguments, supposedly based on pure objective theory, to prove that the armaments on which they chiefly relied were defensive, while those of potential rivals were essentially offensive. Similar attitudes have been taken up in regard to economic " armaments ". In the latter part of the nineteenth century — and in a lesser degree down to 1931 — protective tariffs were commonly regarded in Great Britain as immoral. After 1931 straight tariffs regained their innocence, but barter agreements, industrial (though not

---

[1] R. S. Baker, *Woodrow Wilson and World Settlement*, iii. p. 120. There is an amusing nineteenth-century parallel. " Privateering ", wrote Queen Victoria at the time of the Conference of Paris in 1856, " is a kind of Piracy which disgraces our civilisation ; its abolition throughout the whole world would be a great step in advance." We are not surprised to read that " the privateer was then, like the submarine in modern times, the weapon of the weaker naval Power " (Sir William Malkin, *British Year Book of International Law*, viii. pp. 6, 30).

[2] Published as League of Nations Union Pamphlet No. 76, p. 8. The very word " militarism " conveys to most English readers the same connotation of the peculiar wickedness of armies. It was left to an American historian, Dr. W. L. Langer, to coin the counterpart " navalism ", which has won significantly little acceptance.

agricultural) quotas, exchange controls and other weapons employed by Continental states were still tainted with immorality. Down to 1930, successive revisions of the United States tariff had almost invariably been upward ; and American economists, in other respects staunch upholders of *laissez-faire*, had almost invariably treated tariffs as legitimate and laudable. But the change in the position of the United States from a debtor to a creditor Power, combined with the reversal of British economic policy, altered the picture ; and the reduction of tariff barriers has come to be commonly identified by American spokesmen with the cause of international morality.

National Interest and the Universal Good

The realist should not, however, linger over the infliction of these pin-pricks through chinks in the utopian defences. His task is to bring down the whole cardboard structure of utopian thought by exposing the hollowness of the material out of which it is built. The weapon of the relativity of thought must be used to demolish the utopian concept of a fixed and absolute standard by which policies and actions can be judged. If theories are revealed as a reflexion of practice and principles of political needs, this discovery will apply to the fundamental theories and principles of the utopian creed, and not least to the doctrine of the harmony of interests which is its essential postulate.

It will not be difficult to shew that the utopian, when he preaches the doctrine of the harmony of interests, is innocently and unconsciously adopting Walewski's maxim, and clothing his own interest in the guise of a universal interest for the purpose of imposing it on the rest of the world. " Men come easily to believe that arrangements agreeable to themselves are beneficial to others ", as Dicey observed ; [1] and theories of the public good, which turn out on inspection to be an elegant disguise for some particular interest, are as common in international as in national affairs. The utopian, however eager he may be to establish an absolute standard, does not argue that it is the duty of his country, in conformity with that standard, to put the interest of the world at large before its own interest ; for that would be contrary to his theory that the

[1] Dicey, *Law and Opinion in England* (2nd ed.), pp. 14-15.

interest of all coincides with the interest of each. He argues that what is best for the world is best for his country, and then reverses the argument to read that what is best for his country is best for the world, the two propositions being, from the utopian standpoint, identical ; and this unconscious cynicism of the contemporary utopian has proved a far more effective diplomatic weapon than the deliberate and self-conscious cynicism of a Walewski or a Bismarck. British writers of the past half-century have been particularly eloquent supporters of the theory that the maintenance of British supremacy is the performance of a duty to mankind. " If Great Britain has turned itself into a coal-shed and blacksmith's forge ", remarked *The Times* ingenuously in 1885, " it is for the behoof of mankind as well as its own." [1] The following extract is typical of a dozen which might be culled from memoirs of public men of the period :

> I have but one great object in this world, and that is to maintain the greatness of the Empire. But apart from my John Bull sentiment upon the point, I firmly believe that in doing so I work in the cause of Christianity, of peace, of civilisation, and the happiness of the human race generally. [2]

" I contend that we are the first race in the world," wrote Cecil Rhodes, " and that the more of the world we inhabit the better it is for the human race." [3] In 1891, the most popular and brilliant journalist of the day, W. T. Stead, founded the *Review of Reviews*. " We believe in God, in England and in Humanity ", ran the editorial manifesto in its opening number. " The English-speaking race is one of the chief of God's chosen agents for executing coming improvements in the lot of mankind." [4] An Oxford professor was convinced in 1912 that the secret of Britain's history was that " in fighting for her own independence she has been fighting for the freedom of Europe, and that the service thus rendered to Europe and to mankind has carried with it the possibility of that larger service to which we give the name Empire ".[5]

---

[1] *The Times*, August 27, 1885.
[2] Maurice and Arthur, *The Life of Lord Wolseley*, p. 314.
[3] W. T. Stead, *The Last Will and Testament of Cecil J. Rhodes*, p. 58.
[4] *Review of Reviews*, January 15, 1891.
[5] Spencer Wilkinson, *Government and the War*, p. 116.

The Realist Critique

The first world war carried this conviction to a pitch of emotional frenzy. A bare catalogue, culled from the speeches of British statesmen, of the services which British belligerency was rendering to humanity would fill many pages. In 1917, Balfour told the New York Chamber of Commerce that " since August, 1914, the fight has been for the highest spiritual advantages of mankind, without a petty thought or ambition ".[1] The Peace Conference and its sequel temporarily discredited these professions and threw some passing doubt on the belief in British supremacy as one of the moral assets of mankind. But the period of disillusionment and modesty was short. Moments of international tension, and especially moments when the possibility of war appears on the horizon, always stimulate this identification of national interest with morality. At the height of the Abyssinian crisis, the Archbishop of Canterbury admonished the French public through an interview in a Paris newspaper :

> We are animated by moral and spiritual considerations. I do not think I am departing from my role by contributing towards the clearing up of this misunderstanding. . . .
> It is . . . no egoist interest that is driving us forward, and no consideration of interest should keep you behind.[2]

In the following year, Professor Toynbee was once more able to discover that the security of the British Empire " was also the supreme interest of the whole world ".[3] In 1937, Lord Cecil spoke to the General Council of the League of Nations Union of " our duty to our country, to our Empire and to humanity at large ", and quoted :

> Not once nor twice in our rough island story
> The path of duty is the way to glory.[4]

An Englishman, as Mr. Bernard Shaw remarks in *The Man of Destiny*, " never forgets that the nation which lets its duty get on to the opposite side to its interest is lost ". It is not surprising that an American critic should recently have described the British as " Jesuits lost to the theological but

---

[1] Quoted in Beard, *The Rise of American Civilisation*, ii. p. 646.
[2] Quoted in *Manchester Guardian*, October 18, 1935.
[3] Toynbee, *Survey of International Affairs, 1935*, ii. p. 46.
[4] *Headway*, November 1937.

gained for the political realm ",[1] or that a former Italian Minister for Foreign Affairs should have commented, long before these recent manifestations, on " that precious gift bestowed upon the British people — the possession of writers and clergymen able in perfect good faith to advance the highest moral reasons for the most concrete diplomatic action, with inevitable moral profit to England ".[2]

In recent times, the same phenomenon has become endemic in the United States. The story how McKinley prayed for divine guidance and decided to annex the Philippines is a classic of modern American history ; and this annexation was the occasion of a popular outburst of moral self-approval hitherto more familiar in the foreign policy of Great Britain than of the United States. Theodore Roosevelt, who believed more firmly than any previous American President in the doctrine *L'état, c'est moi*, carried the process a step further. The following curious dialogue occurred in his cross-examination during a libel action brought against him in 1915 by a Tammany leader :

Query : How did you know that substantial justice was done ?

ROOSEVELT : Because I did it, because . . . I was doing my best.

Query : You mean to say that, when you do a thing, thereby substantial justice is done.

ROOSEVELT : I do. When I do a thing, I do it so as to do substantial justice. I mean just that.[3]

Woodrow Wilson was less naively egotistical, but more profoundly confident of the identity of American policy and universal justice. After the bombardment of Vera Cruz in 1914, he assured the world that " the United States had gone down to Mexico to serve mankind ".[4] During the first world war, he advised American naval cadets " not only always to think first of America, but always, also, to think first of humanity " — a feat rendered slightly less difficult by his explanation that the United States had been " founded for the benefit of

[1] Carl Becker, *Yale Review*, xxvii. p. 452.
[2] Count Sforza, *Foreign Affairs*, October 1927, p. 67.
[3] Quoted in H. F. Pringle, *Theodore Roosevelt*, p. 318.
[4] *Public Papers of Woodrow Wilson : The New Democracy*, ed. R. S. Baker, i. p. 104.

humanity ".[1] Shortly before the entry of the United States into the war, in an address to the Senate on war aims, he stated the identification still more categorically : " These are American principles, American policies. . . . They are the principles of mankind and must prevail." [2]

It will be observed that utterances of this character proceed almost exclusively from Anglo-Saxon statesmen and writers. It is true that when a prominent National Socialist asserted that " anything that benefits the German people is right, anything that harms the German people is wrong ",[3] he was merely propounding the same identification of national interest with universal right which had already been established for English-speaking countries by Wilson, Professor Toynbee, Lord Cecil and many others. But when the claim is translated into a foreign language, the note seems forced, and the identification unconvincing, even to the peoples concerned. Two explanations are commonly given of this curious discrepancy. The first explanation, which is popular in English-speaking countries, is that the policies of the English-speaking nations are in fact more virtuous and disinterested than those of Continental states, so that Wilson and Professor Toynbee and Lord Cecil are, broadly speaking, right when they identify the American and British national interests with the interest of mankind. The second explanation, which is popular in Continental countries, is that the English-speaking peoples are past masters in the art of concealing their selfish national interests in the guise of the general good, and that this kind of hypocrisy is a special and characteristic peculiarity of the Anglo-Saxon mind.

It seems unnecessary to accept either of these heroic attempts to cut the knot. The solution is a simple one. Theories of social morality are always the product of a dominant group which identifies itself with the community as a whole, and which possesses facilities denied to subordinate groups or individuals for imposing its view of life on the community. Theories of international morality are, for the same reason and in virtue of the same process, the product of dominant nations or groups of nations. For the past hundred years, and more especially since 1918, the English-speaking peoples have formed

---

[1] *Public Papers of Woodrow Wilson: The New Democracy*, ed. R. S. Baker, i. pp. 318-19.   [2] *Ibid.* ii. p. 414.

[3] Quoted in Toynbee, *Survey of International Affairs, 1936*, p. 319.

the dominant group in the world ; and current theories of international morality have been designed to perpetuate their supremacy and expressed in the idiom peculiar to them. France, retaining something of her eighteenth-century tradition and restored to a position of dominance for a short period after 1918, has played a minor part in the creation of current international morality, mainly through her insistence on the role of law in the moral order. Germany, never a dominant Power and reduced to helplessness after 1918, has remained for these reasons outside the charmed circle of creators of international morality. Both the view that the English-speaking peoples are monopolists of international morality and the view that they are consummate international hypocrites may be reduced to the plain fact that the current canons of international virtue have, by a natural and inevitable process, been mainly created by them.

The Realist Critique of the Harmony of Interests

The doctrine of the harmony of interests yields readily to analysis in terms of this principle. It is the natural assumption of a prosperous and privileged class, whose members have a dominant voice in the community and are therefore naturally prone to identify its interest with their own. In virtue of this identification, any assailant of the interests of the dominant group is made to incur the odium of assailing the alleged common interest of the whole community, and is told that in making this assault he is attacking his own higher interests. The doctrine of the harmony of interests thus serves as an ingenious moral device invoked, in perfect sincerity, by privileged groups in order to justify and maintain their dominant position. But a further point requires notice. The supremacy within the community of the privileged group may be, and often is, so overwhelming that there is, in fact, a sense in which its interests are those of the community, since its well-being necessarily carries with it some measure of well-being for other members of the community, and its collapse would entail the collapse of the community as a whole. In so far, therefore, as the alleged natural harmony of interests has any reality, it is created by the overwhelming power of the privileged group, and is an excellent illustration of the Machiavellian maxim that morality is the product of power. A few examples will make this analysis of the doctrine of the harmony of interests clear.

In the nineteenth century, the British manufacturer or merchant, having discovered that *laissez-faire* promoted his own prosperity, was sincerely convinced that it also promoted British prosperity as a whole. Nor was this alleged harmony of interests between himself and the community entirely fictitious. The predominance of the manufacturer and the merchant was so overwhelming that there was a sense in which an identity between their prosperity and British prosperity as a whole could be correctly asserted. From this it was only a short step to argue that a worker on strike, in damaging the prosperity of the British manufacturer, was damaging British prosperity as a whole, and thereby damaging his own, so that he could be plausibly denounced by the predecessors of Professor Toynbee as immoral and by the predecessors of Professor Zimmern as muddle-headed. Moreover, there was a sense in which this argument was perfectly correct. Nevertheless, the doctrine of the harmony of interests and of solidarity between the classes must have seemed a bitter mockery to the under-privileged worker, whose inferior status and insignificant stake in " British prosperity " were consecrated by it ; and presently he was strong enough to force the abandonment of *laissez-faire* and the substitution for it of the " social service state ", which implicitly denies the natural harmony of interests and sets out to create a new harmony by artificial means.

The same analysis may be applied in international relations. British nineteenth-century statesmen, having discovered that free trade promoted British prosperity, were sincerely convinced that, in doing so, it also promoted the prosperity of the world as a whole. British predominance in world trade was at that time so overwhelming that there was a certain undeniable harmony between British interests and the interests of the world. British prosperity flowed over into other countries, and a British economic collapse would have meant world-wide ruin. British free traders could and did argue that protectionist countries were not only egotistically damaging the prosperity of the world as a whole, but were stupidly damaging their own, so that their behaviour was both immoral and muddle headed. In British eyes, it was irrefutably proved that international trade was a single whole, and flourished or slumped together. Nevertheless, this alleged international harmony of interests seemed a mockery to those under-privileged nations whose inferior status and insignificant stake in international trade were consecrated by it. The revolt against it destroyed that overwhelming British preponderance which had provided a plausible basis for the theory. Economically, Great Britain in the nineteenth century was dominant enough to make a bold bid to impose on the world her own conception of international economic morality. When competition of all against all replaced the domination of the world market by a single Power, conceptions of international economic morality necessarily became chaotic.

Politically, the alleged community of interest in the maintenance of peace, whose ambiguous character has already been discussed, is capitalised in the same way by a dominant nation or group of nations. Just as the ruling class in a community prays for domestic peace, which guarantees its own security and predominance, and denounces class-war, which might threaten them, so international peace becomes a special vested interest of predominant Powers. In the past, Roman and British imperialism were commended to the world in the guise of the *pax Romana* and the *pax Britannica*. To-day, when no single Power is strong enough to dominate the world, and supremacy is vested in a group of nations, slogans like " collective security " and " resistance to aggression " serve the same purpose of proclaiming an identity of interest between the dominant group and the world as a whole in the maintenance of peace. Moreover, as in the examples we have just considered, so long as the supremacy of the dominant group is sufficiently great, there is a sense in which this identity of interests exists. " England ", wrote a German professor in the nineteen-twenties, " is the solitary Power with a national programme which, while egotistic through and through, at the same time promises to the world something which the world passionately desires : order, progress and eternal peace." [1] When Mr. Churchill declared that " the fortunes of the British Empire and its glory are inseparably interwoven with the fortunes of the world ",[2] this statement had precisely the same

---

[1] Dibelius, *England*, p. 109.
[2] Winston Churchill, *Arms and the Covenant*, p 272.

foundation in fact as the statement that the prosperity of British manufacturers in the nineteenth century was inseparably interwoven with British prosperity as a whole. Moreover, the purpose of the statements was precisely the same, namely to establish the principle that the defence of the British Empire, or the prosperity of the British manufacturer, was a matter of common interest to the whole community, and that anyone who attacked it was therefore either immoral or muddle-headed. It is a familiar tactic of the privileged to throw moral discredit on the under-privileged by depicting them as disturbers of the peace ; and this tactic is as readily applied internationally as within the national community. " International law and order ", writes Professor Toynbee of a recent crisis, " were in the true interests of the whole of mankind . . . whereas the desire to perpetuate the region of violence in international affairs was an anti-social desire which was not even in the ultimate interests of the citizens of the handful of states that officially professed this benighted and anachronistic creed." [1] This is precisely the argument, compounded of platitude and falsehood in about equal parts, which did duty in every strike in the early days of the British and American Labour movements. It was common form for employers, supported by the whole capitalist press, to denounce the " anti-social " attitude of trade union leaders, to accuse them of attacking law and order and of introducing " the reign of violence ", and to declare that " true " and " ultimate " interests of the workers lay in peaceful co-operation with the employers. [2] In the field of social relations, the disingenuous character of this argument has long been recognised. But just as the threat of class-war by the proletarian is " a natural cynical reaction to the sentimental and dishonest efforts of the privileged classes to obscure the conflict of interest between classes by a constant emphasis on the minimum interests which they have in common ", [3] so the war-mongering of the dissatisfied Powers was the " natural, cynical reaction " to the sentimental and dishonest platitudinis-

---

[1] Toynbee, *Survey of International Affairs, 1935,* ii. p. 46.

[2] " Pray earnestly that right may triumph ", said the representative of the Philadelphia coal-owners in an early strike organised by the United Mine Workers, " remembering that the Lord God Omnipotent still reigns, and that His reign is one of law and order, and not of violence and crime " (H. F. Pringle, *Theodore Roosevelt*, p. 267).

[3] R. Niebuhr, *Moral Man and Immoral Society*, p. 153.

ing of the satisfied Powers on the common interest in peace. When Hitler refused to believe " that God has permitted some nations first to acquire a world by force and then to defend this robbery with moralising theories ",[1] he was merely echoing in another context the Marxist denial of a community of interest between " haves " and " have-nots ", the Marxist exposure of the interested character of " *bourgeois* morality ", and the Marxist demand for the expropriation of the expropriators.

The crisis of September 1938 demonstrated in a striking way the political implications of the assertion of a common interest in peace. When Briand proclaimed that " peace comes before all ", or Mr. Eden that " there is no dispute which cannot be settled by peaceful means ",[2] the assumption underlying these platitudes was that, so long as peace was maintained, no changes distasteful to France or Great Britain could be made in the *status quo*. In 1938, France and Great Britain were trapped by the slogans which they themselves had used in the past to discredit the dissatisfied Powers, and Germany had become sufficiently dominant (as France and Great Britain had hitherto been) to turn the desire for peace to her own advantage. About this time, a significant change occurred in the attitude of the German and Italian dictators. Hitler eagerly depicted Germany as a bulwark of peace menaced by war-mongering democracies. The League of Nations, he declared in his Reichstag speech of April 28, 1939, is a " stirrer up of trouble ", and collective security means " continuous danger of war ". Mussolini borrowed the British formula about the possibility of settling all international disputes by peaceful means, and declared that " there are not in Europe at present problems so big and so active as to justify a war which from a European conflict would naturally become universal ".[3] Such utterances were symptoms that Germany and Italy were already looking forward to the time when, as dominant Powers, they would acquire the vested interest in peace recently enjoyed by Great Britain and France, and be able to get their way by pillorying the democratic countries as enemies of peace. These developments may have made it easier to appreciate Halévy's

[1] Speech in the Reichstag, January 30, 1939.
[2] *League of Nations : Eighteenth Assembly*, p. 63.
[3] *The Times*, May 15, 1939.

subtle observation that " propaganda against war is itself a form of war propaganda ".[1]

The Realist Critique of Internationalism

The concept of internationalism is a special form of the doctrine of the harmony of interests. It yields to the same analysis ; and there are the same difficulties about regarding it as an absolute standard independent of the interests and policies of those who promulgate it. " Cosmopolitanism ", wrote Sun Yat-sen, " is the same thing as China's theory of world empire two thousand years ago. . . . China once wanted to be sovereign lord of the earth and to stand above every other nation, so she espoused cosmopolitanism." [2] In the Egypt of the Eighteenth Dynasty, according to Freud, " imperialism was reflected in religion as universality and monotheism ".[3] The doctrine of a single world-state, propagated by the Roman Empire and later by the Catholic Church, was the symbol of a claim to universal dominion. Modern internationalism has its genesis in seventeenth- and eighteenth-century France, during which French hegemony in Europe was at its height. This was the period which produced Sully's *Grand Dessin* and the Abbé Saint-Pierre's *Projet de Paix Perpétuelle* (both plans to perpetuate an international *status quo* favourable to the French monarchy), which saw the birth of the humanitarian and cosmopolitan doctrines of the Enlightenment, and which established French as the universal language of educated people. In the next century, the leadership passed to Great Britain, which became the home of internationalism. On the eve of the Great Exhibition of 1851 which, more than any other single event, established Great Britain's title to world supremacy, the Prince Consort spoke movingly of " that great end to which . . . all history points — the realisation of the unity of mankind " ; [4] and Tennyson hymned " the parliament of man, the federation of the world ". France chose the moment of her greatest supremacy in the nineteen-twenties to launch a plan of " European Union " ; and Japan shortly afterwards developed an

[1] Halévy, *A History of the English People in 1895-1905* (Engl. transl.), i. Introduction, p. xi.
[2] Sun Yat-sen, *San Min Chu I* (Engl. transl.), pp. 68-9.
[3] Sigmund Freud, *Moses and Monotheism*, p. 36.
[4] T Martin, *Life of the Prince Consort*, iii. p. 247.

ambition to proclaim herself the leader of a united Asia. It was symptomatic of the growing international predominance of the United States when widespread popularity was enjoyed in the late nineteen-thirties by the book of an American journalist advocating a world union of democracies, in which the United States would play the predominant role.[1]

Just as pleas for " national solidarity " in domestic politics always come from a dominant group which can use this solidarity to strengthen its own control over the nation as a whole, so pleas for international solidarity and world union come from those dominant nations which may hope to exercise control over a unified world. Countries which are struggling to force their way into the dominant group naturally tend to invoke nationalism against the internationalism of the controlling Powers. In the sixteenth century, England opposed her nascent nationalism to the internationalism of the Papacy and the Empire. In the past century and a half Germany opposed her nascent nationalism to the internationalism first of France, then of Great Britain. This circumstance made her impervious to those universalist and humanitarian doctrines which were popular in eighteenth-century France and nineteenth-century Britain ; and her hostility to internationalism was further aggravated after 1919, when Great Britain and France endeavoured to create a new " international order " as a bulwark of their own predominance. " By ' international '," wrote a German correspondent in *The Times*, " we have come to understand a conception that places other nations at an advantage over our own." [2] Nevertheless, there was little doubt that Germany, if she became supreme in Europe, would adopt international slogans and establish some kind of international organisation to bolster up her power. A British Labour ex-Minister at one moment advocated the suppression of Article 16 of the Covenant of the League of Nations on the unexpected ground that the totalitarian states might some day capture the League and invoke that article to justify the use of force by themselves.[3] It seemed more likely that they would seek to develop the Anti-Comintern Pact into some form of

[1] Clarence Streit, *Union Now*.
[2] *The Times*, November 5, 1938.
[3] Lord Marley in the House of Lords, November 30, 1938 : *Official Report*, col. 258.

international organisation " The Anti-Comintern Pact ", said
Hitler in the Reichstag on January 30, 1939, " will perhaps
one day become the crystallisation point of a group of Powers
whose ultimate aim is none other than to eliminate the menace
to the peace and culture of the world instigated by a satanic
apparition." " Either Europe must achieve solidarity," remarked an Italian journal about the same time, " or the ' axis '
will impose it." [1] " Europe in its entirety ", said Goebbels,
" is adopting a new order and a new orientation under the
intellectual leadership of National Socialist Germany and
Fascist Italy." [2] These were symptoms not of a change of heart,
but of the fact that Germany and Italy felt themselves to be
approaching the time when they might become strong enough to
espouse internationalism. " International order " and " international solidarity " will always be slogans of those who feel
strong enough to impose them on others.

The exposure of the real basis of the professedly abstract
principles commonly invoked in international politics is the
most damning and most convincing part of the realist indict-
ment of utopianism. The nature of the charge is frequently
misunderstood by those who seek to refute it. The charge is
not that human beings fail to live up to their principles. It
matters little that Wilson, who thought that the right was more
precious than peace, and Briand, who thought that peace came
even before justice, and Mr. Eden, who believed in collective
security, failed themselves, or failed to induce their countrymen,
to apply these principles consistently. What matters is that
these supposedly absolute and universal principles were not
principles at all, but the unconscious reflexions of national
policy based on a particular interpretation of national interest
at a particular time. There is a sense in which peace and
co-operation between nations or classes or individuals is a
common and universal end irrespective of conflicting interests
and politics. There is a sense in which a common interest
exists in the maintenance of order, whether it be international
order or " law and order " within the nation. But as soon as
the attempt is made to apply these supposedly abstract principles to a concrete political situation, they are revealed as the

---

[1] *Relazioni Internazionali*, quoted in *The Times*, December 5, 1938.
[2] *Völkischer Beobachter*, April 1, 1939.

transparent disguises of selfish vested interests. The bankruptcy of utopianism resides not in its failure to live up to its principles, but in the exposure of its inability to provide any absolute and disinterested standard for the conduct of international affairs. The utopian, faced by the collapse of standards whose interested character he has failed to penetrate, takes refuge in condemnation of a reality which refuses to conform to these standards. A passage penned by the German historian Meinecke after the first world war is the best judgment by anticipation of the role of utopianism in the international politics of the period :

> The profound defect of the Western, natural-law type of thought was that, when applied to the real life of the state, it remained a dead letter, did not penetrate the consciousness of statesmen, did not hinder the modern hypertrophy of state interest, and so led either to aimless complaints and doctrinaire suppositions or else to inner falsehood and cant.[1]

These " aimless complaints ", these " doctrinaire suppositions ", this " inner falsehood and cant " will be familiar to all those who have studied what was written about international politics in English-speaking countries between the two world wars.

[1] Meinecke, *Staatsräson*, p. 533.

CHAPTER 6

THE LIMITATIONS OF REALISM

THE exposure by realist criticism of the hollowness of the utopian edifice is the first task of the political thinker. It is only when the sham has been demolished that there can be any hope of raising a more solid structure in its place. But we cannot ultimately find a resting place in pure realism ; for realism, though logically overwhelming, does not provide us with the springs of action which are necessary even to the pursuit of thought. Indeed, realism itself, if we attack it with its own weapons, often turns out in practice to be just as much conditioned as any other mode of thought. In politics, the belief that certain facts are unalterable or certain trends irresistible commonly reflects a lack of desire or lack of interest to change or resist them. The impossibility of being a consistent and thorough-going realist is one of the most certain and most curious lessons of political science. Consistent realism excludes four things which appear to be essential ingredients of all effective political thinking : a finite goal, an emotional appeal, a right of moral judgment and a ground for action.

The conception of politics as an infinite process seems in the long run uncongenial or incomprehensible to the human mind. Every political thinker who wishes to make an appeal to his contemporaries is consciously or unconsciously led to posit a finite goal. Treitschke declared that the " terrible thing " about Machiavelli's teaching was " not the immorality of the methods he recommends, but the lack of content of the state, which exists only in order to exist ".[1] In fact, Machiavelli is not so consistent. His realism breaks down in the last chapter of *The Prince*, which is entitled " An Exhortation to free Italy from the Barbarians " — a goal whose necessity could be deduced from no realist premise. Marx, having dissolved human thought and action into the relativism of the dialectic, postulates the absolute goal of a classless society where the dialectic no longer operates — that one far-off event towards which, in true Victorian fashion, he believed the whole creation

[1] Treitschke, *Aufsätze*, iv. p. 428.

to be moving. The realist thus ends by negating his own postulate and assuming an ultimate reality outside the historical process. Engels was one of the first to level this charge against Hegel. " The whole dogmatic content of the Hegelian system is declared to be absolute truth in contradiction to his dialectical method, which dissolves all dogmatism." [1] But Marx lays himself open to precisely the same criticism when he brings the process of dialectical materialism to an end with the victory of the proletariat. Thus utopianism penetrates the citadel of realism ; and to envisage a continuing, but not infinite, process towards a finite goal is shewn to be a condition of political thought. The greater the emotional stress, the nearer and more concrete is the goal. The first world war was rendered tolerable by the belief that it was the last of wars. Woodrow Wilson's moral authority was built up on the conviction, shared by himself, that he possessed the key to a just, comprehensive and final settlement of the political ills of mankind. It is noteworthy that almost all religions agree in postulating an ultimate state of complete blessedness.

The finite goal, assuming the character of an apocalyptic vision, thereby acquires an emotional, irrational appeal which realism itself cannot justify or explain. Everyone knows Marx's famous prediction of the future classless paradise :

> When work ceases to be merely a means of life and becomes the first living need ; when, with the all-round development of the individual, productive forces also develop, and all the sources of collective wealth flow in free abundance — then only will it be possible to transcend completely the narrow horizon of *bourgeois* right, and society can inscribe on its banner : From each according to his capacities, to each according to his needs. [2]

Sorel proclaimed the necessity of a " myth " to make revolutionary teaching effective ; and Soviet Russia has exploited for this purpose the myth, first of world revolution, and more recently of the " socialist fatherland ". There is much to be said for Professor Laski's view that " communism has made its way by its idealism, and not by its realism, by its spiritual promise, not by its materialistic prospects ". [3] A modern theo-

[1] Engels, *Ludwig Feuerbach* (Engl. transl.), p. 23.
[2] Marx and Engels, *Works* (Russian ed.), xv. p. 275.
[3] Laski, *Communism*, p. 250.

logian has analysed the situation with almost cynical clear-
sightedness :

> Without the ultrarational hopes and passions of religion,
> no society will have the courage to conquer despair and
> attempt the impossible ; for the vision of a just society is
> an impossible one, which can be approximated only by those
> who do not regard it as impossible. The truest visions of
> religion are illusions, which may be partly realised by being
> resolutely believed.[1]

And this again closely echoes a passage in *Mein Kampf* in
which Hitler contrasts the " programme-maker " with the
politician :

> His [i.e. the programme-maker's] significance lies almost
> wholly in the future, and he is often what one means by the
> word '*weltfremd*' [unpractical, utopian]. For if the art of
> the politician is really the art of the possible, then the pro-
> gramme-maker belongs to those of whom it is said that they
> please the gods only if they ask and demand from them the
> impossible.[2]

Credo quia impossibile becomes a category of political thinking.

Consistent realism, as has already been noted, involves
acceptance of the whole historical process and precludes moral
judgments on it. As we have seen, men are generally prepared
to accept the judgment of history on the past, praising success
and condemning failure. This test is also widely applied to
contemporary politics. Such institutions as the League of
Nations, or the Soviet or Fascist régimes, are to a considerable
extent judged by their capacity to achieve what they profess
to achieve ; and the legitimacy of this test is implicitly admitted
by their own propaganda, which constantly seeks to exaggerate
their successes and minimise their failures. Yet it is clear that
mankind as a whole is not prepared to accept this rational
test as a universally valid basis of political judgment. The
belief that whatever succeeds is right, and has only to be
understood to be approved, must, if consistently held, empty
thought of purpose, and thereby sterilise and ultimately destroy
it. Nor do those whose philosophy appears to exclude the
possibility of moral judgments in fact refrain from pronouncing

---

[1] R. Niebuhr, *Moral Man and Immoral Society*, p. 81.

[2] Hitler, *Mein Kampf*, p. 231.

them. Frederick the Great, having explained that treaties should be observed for the reason that " one can trick only once ", goes on to call the breaking of treaties " a bad and knavish policy ", though there is nothing in his thesis to justify the moral epithet.[1] Marx, whose philosophy appeared to demonstrate that capitalists could only act in a certain way, spends many pages — some of the most effective in *Capital* — in denouncing the wickedness of capitalists for behaving in precisely that way. The necessity, recognised by all politicians, both in domestic and in international affairs, for cloaking interests in a guise of moral principles is in itself a symptom of the inadequacy of realism. Every age claims the right to create its own values, and to pass judgments in the light of them ; and even if it uses realist weapons to dissolve other values, it still believes in the absolute character of its own. It refuses to accept the implication of realism that the word " ought " is meaningless.

Most of all, consistent realism breaks down because it fails to provide any ground for purposive or meaningful action. If the sequence of cause and effect is sufficiently rigid to permit of the " scientific prediction " of events, if our thought is irrevocably conditioned by our status and our interests, then both action and thought become devoid of purpose. If, as Schopenhauer maintains, " the true philosophy of history consists of the insight that, throughout the jumble of all these ceaseless changes, we have ever before our eyes the same unchanging being, pursuing the same course to-day, yesterday and for ever ",[2] then passive contemplation is all that remains to the individual. Such a conclusion is plainly repugnant to the most deep-seated belief of man about himself. That human affairs can be directed and modified by human action and human thought is a postulate so fundamental that its rejection seems scarcely compatible with existence as a human being. Nor is it in fact rejected by those realists who have left their mark on history. Machiavelli, when he exhorted his compatriots to be good Italians, clearly assumed that they were free to follow or ignore his advice. Marx, by birth and training a *bourgeois*, believed himself free to think and act like a proletarian, and regarded it as his mission to persuade others,

[1] *Anti-Machiavel*, p. 248.
[2] Schopenhauer, *Welt als Wille und Vorstellung*, ii. ch. 38.

whom he assumed to be equally free, to think and act likewise. Lenin, who wrote of the imminence of world revolution as a " scientific prediction ", admitted elsewhere that " no situations exist from which there is absolutely no way out ".[1] In moments of crisis, Lenin appealed to his followers in terms which might equally well have been used by so thorough-going a believer in the power of the human will as Mussolini or by any other leader of any period : " At the decisive moment and in the decisive place, you *must prove* the stronger, you must *be victorious* ".[2] Every realist, whatever his professions, is ultimately compelled to believe not only that there is something which man ought to think and do, but that there is something which he can think and do, and that his thought and action are neither mechanical nor meaningless.

We return therefore to the conclusion that any sound political thought must be based on elements of both utopia and reality. Where utopianism has become a hollow and intolerable sham, which serves merely as a disguise for the interests of the privileged, the realist performs an indispensable service in unmasking it. But pure realism can offer nothing but a naked struggle for power which makes any kind of international society impossible. Having demolished the current utopia with the weapons of realism, we still need to build a new utopia of our own, which will one day fall to the same weapons. The human will will continue to seek an escape from the logical consequences of realism in the vision of an international order which, as soon as it crystallises itself into concrete political form, becomes tainted with self-interest and hypocrisy, and must once more be attacked with the instruments of realism.

Here, then, is the complexity, the fascination and the tragedy of all political life. Politics are made up of two elements — utopia and reality — belonging to two different planes which can never meet. There is no greater barrier to clear political thinking than failure to distinguish between ideals, which are utopia, and institutions, which are reality. The communist who set communism against democracy was usually thinking of communism as a pure ideal of equality and brotherhood, and of democracy as an institution which existed in Great Britain, France or the United States and which exhibited the

[1] Lenin, *Works* (2nd Russian ed.), xxv. p. 340.
[2] Lenin, *Collected Works* (Engl. transl.), xxi. pt. i. p. 68.

vested interests, the inequalities and the oppression inherent in all political institutions. The democrat who made the same comparison was in fact comparing an ideal pattern of democracy laid up in heaven with communism as an institution existing in Soviet Russia with its class-divisions, its heresy-hunts and its concentration camps. The comparison, made in each case between an ideal and an institution, is irrelevant and makes no sense. The ideal, once it is embodied in an institution, ceases to be an ideal and becomes the expression of a selfish interest, which must be destroyed in the name of a new ideal. This constant interaction of irreconcileable forces is the stuff of politics. Every political situation contains mutually incompatible elements of utopia and reality, of morality and power.

This point will emerge more clearly from the analysis of the nature of politics which we have now to undertake.

PART THREE

POLITICS, POWER AND MORALITY

CHAPTER 7

THE NATURE OF POLITICS

MAN has always lived in groups. The smallest kind of human group, the family, has clearly been necessary for the maintenance of the species. But so far as is known, men have always from the most primitive times formed semi-permanent groups larger than the single family; and one of the functions of such a group has been to regulate relations between its members. Politics deals with the behaviour of men in such organised permanent or semi-permanent groups. All attempts to deduce the nature of society from the supposed behaviour of man in isolation are purely theoretical, since there is no reason to assume that such a man ever existed. Aristotle laid the foundation of all sound thinking about politics when he declared that man was by nature a political animal.

Man in society reacts to his fellow men in two opposite ways. Sometimes he displays egoism, or the will to assert himself at the expense of others. At other times he displays sociability, or the desire to co-operate with others, to enter into reciprocal relations of good-will and friendship with them, and even to subordinate himself to them. In every society, these two qualities can be seen at work. No society can exist unless a substantial proportion of its members exhibits in some degree the desire for co-operation and mutual good-will. But in every society some sanction is required to produce the measure of solidarity requisite for its maintenance; and this sanction is applied by a controlling group or individual acting in the name of the society. Membership of most societies is voluntary, and the only ultimate sanction which can be applied is expulsion. But the peculiarity of political society, which in the modern world takes the form of the state, is that membership is compulsory. The state, like other societies, must be based on some sense of common interests and obligations among its members. But coercion is regularly exercised by a governing group to enforce loyalty and obedience ; and this coercion inevitably means that the governors control the governed and " exploit " them for their own purposes.[1]

The dual character of political society is therefore strongly marked. Professor Laski tells us that " every state is built upon the consciences of men ".[2] On the other hand, anthropology, as well as much recent history, teaches that " war seems to be the main agency in producing the state " ; [3] and Professor Laski himself, in another passage, declares that " our civilisation is held together by fear rather than by good-will ".[4] There is no contradiction between these apparently opposite views. When Tom Paine, in the *Rights of Man*, tries to confront Burke with the dilemma that " governments arise either *out* of the people or *over* the people ", the answer is that they do both. Coercion and conscience, enmity and good-will, self-assertion and self-subordination, are present in every political society. The state is built up out of these two conflicting aspects of human nature. Utopia and reality, the ideal and the institution, morality and power, are from the outset inextricably blended in it. In the making of the United States, as a modern American writer has said, " Hamilton stood for strength, wealth and power, Jefferson for the American dream "; and both the power and the dream were necessary ingredients.[5]

If this be correct, we can draw one important conclusion.

[1] " Everywhere do I perceive a certain conspiracy of the rich men seeking their own advantage under the name and pretext of the commonwealth " (More, *Utopia*). " The exploitation of one part of society by another is common to all past centuries " (*Communist Manifesto*).

[2] *A Defence of Liberty against Tyrants* (*Vindiciae contra Tyrannos*), ed. Laski, Introd. p. 55.   [3] Linton, *The Study of Man*, p. 240.

[4] Laski, *A Grammar of Politics*, p. 20.

[5] J. Truslow Adams, *The Epic of America*, p. 112. The idea that the state has a moral foundation in the consent of its citizens as well as a power foundation was propounded by Locke and Rousseau and popularised by the American and French revolutions. Two recent expressions of the idea may be quoted. The Czecho-Slovak declaration of independence of October 18, 1918, described Austria-Hungary as " a state which has no justification for its existence, and which, since it refuses to accept the fundamental basis of modern world-organisation [i.e. self-determination], is only an artificial and unmoral construction ". In February 1938, Hitler told Schuschnigg, the then Austrian Chancellor, that " a régime lacking every kind of legality and which in reality ruled only by force, must in the long run come into continually increasing conflict with public opinion " (speech in the Reichstag of March 17, 1938). Hitler maintained that the two pillars of the state are " force " and " popularity " (*Mein Kampf*, p. 579).

The utopian who dreams that it is possible to eliminate self-assertion from politics and to base a political system on morality alone is just as wide of the mark as the realist who believes that altruism is an illusion and that all political action is based on self-seeking. These errors have both left their mark on popular terminology. The phrase " power politics " is often used in an invidious sense, as if the element of power or self-assertion in politics were something abnormal and susceptible of elimination from a healthy political life. Conversely, there is a disposition, even among some writers who are not strictly speaking realists, to treat politics as the science of power and self-assertion and exclude from it by definition actions inspired by the moral consciousness. Professor Catlin describes the *homo politicus* as one who " seeks to bring into conformity with his own will the wills of others, so that he may the better attain his own ends ".[1] Such terminological implications are misleading. Politics cannot be divorced from power. But the *homo politicus* who pursues nothing but power is as unreal a myth as the *homo economicus* who pursues nothing but gain. Political action must be based on a co-ordination of morality and power.

This truth is of practical as well as theoretical importance. It is as fatal in politics to ignore power as it is to ignore morality. The fate of China in the nineteenth century is an illustration of what happens to a country which is content to believe in the moral superiority of its own civilisation and to despise the ways of power. The Liberal Government of Great Britain nearly came to grief in the spring of 1914 because it sought to pursue an Irish policy based on moral authority unsupported (or rather, directly opposed) by effective military power. In Germany, the Frankfort Assembly of 1848 is the classic example of the impotence of ideas divorced from power ; and the Weimar Republic broke down because many of the policies it pursued — in fact, nearly all of them except its opposition to the communists — were unsupported, or actively opposed, by effective military power.[2] The utopian, who believes that democracy is not based on force, refuses to look these unwelcome facts in the face.

[1] Catlin, *The Science and Method of Politics*, p. 309.

[2] It is significant that the word *Realpolitik* was coined in the once famous treatise of von Rochau, *Grundsätze der Realpolitik* published in 1853, which was largely inspired by the lessons of Frankfort. The inspiration which Hitler's *Realpolitik* has derived from the lessons of the Weimar Republic is obvious.

On the other hand, the realist, who believes that, if you look after the power, the moral authority will look after itself, is equally in error. The most recent form of this doctrine is embodied in the much-quoted phrase : " The function of force is to give moral ideas time to take root ". Internationally, this argument was used in 1919 by those who, unable to defend the Versailles Treaty on moral grounds, maintained that this initial act of power would pave the way for subsequent moral appeasement. Experience has done little to confirm this comfortable belief. The same fallacy is implicit in the once popular view that the aim of British policy should be " to rebuild the League of Nations, to make it capable of holding a political aggressor in restraint by armed power, and thereafter to labour faithfully for the mitigation of just and real grievances ".[1] Once the enemy has been crushed or the " aggressor " restrained by force, the " thereafter " fails to arrive. The illusion that priority can be given to power and that morality will follow, is just as dangerous as the illusion that priority can be given to moral authority and that power will follow.

Before proceeding, however, to consider the respective roles of power and morality in politics, we must take some note of the views of those who, though far from being realists, identify politics with power and believe that moral concepts must be altogether excluded from its scope. There is, according to this view, an essential antinomy between politics and morality ; and the moral man as such will therefore have nothing to do with politics. This thesis has many attractions, and reappears at different periods of history and in different contexts. It takes at least three forms.

(i) Its simplest form is the doctrine of non-resistance. The moral man recognises the existence of political power as an evil, but regards the use of power to resist power as a still greater evil. This is the basis of such doctrines of non-resistance as those of Jesus or of Gandhi, or of modern pacifism. It amounts, in brief, to a boycott of politics.

(ii) The second form of the antithesis between politics and morality is anarchism. The state, as the principal organ of political power, is " the most flagrant, most cynical and most

---

[1] Winston Churchill, *Arms and the Covenant*, p. 368. The argument that power is a necessary motive force for the remedy of " just " grievances is further developed on pp. 209-216.

complete negation of humanity ".[1] The anarchist will use power to overthrow the state. This revolutionary power is, however, not thought of as political power, but as the spontaneous revolt of the outraged individual conscience. It does not seek to create a new political society to take the place of the old one, but a moral society from which power, and consequently politics, are completely eliminated. " The principles of the Sermon on the Mount ", an English divine recently remarked, would mean " sudden death to civilised society." [2] The anarchist sets out to destroy " civilised society " in the name of the Sermon on the Mount.

(iii) A third school of thought starts from the same premise of the essential antithesis between morality and politics, but arrives at a totally different conclusion. The injunction of Jesus to " render unto Caesar the things that are Caesar's, and unto God the things that are God's ", implies the coexistence of two separate spheres : the political and the moral. But the moral man is under an obligation to assist — or at any rate not to obstruct — the politician in the discharge of his non-moral functions. " Let every soul be subject to the higher powers. The powers that be are ordained of God." We thus recognise politics as necessary but non-moral. This tradition, which remained dormant throughout the Middle Ages, when the ecclesiastical and the secular authority was theoretically one, was revived by Luther in order to effect his compromise between reformed church and state. Luther " turned on the peasants of his day in holy horror when they attempted to transmute the 'spiritual' kingdom into an 'earthly' one by suggesting that the principles of the gospel had social significance ".[3] The division of functions between Caesar and God is implicit in the very conception of an " established " church. But the tradition has been more persistent and more effective in Lutheran Germany than anywhere else. " We do not consult Jesus ", wrote a German liberal nineteenth-century pastor, " when we are concerned with things which belong to the domain of the construction of the state and political economy " ;[4] and Bernhardi declared that "Christian

---

[1] Bakunin, *Œuvres*, i. p 150 ; cf. vi. p. 17 : " If there is a devil in all human history, it is this principle of command and authority ".

[2] The Dean of St. Paul's, quoted in a leading article in *The Times*, August 2, 1937.   [3] R. Niebuhr, *Moral Man and Immoral Society*, p. 77.

[4] Quoted in W. F. Bruck, *Social and Economic History of Germany*, p. 65.

morality is personal and social, and in its nature cannot be political ".[1] The same attitude is inherent in the modern theology of Karl Barth, which insists that political and social evils are the necessary product of man's sinful nature and that human effort to eradicate them is therefore futile ; and the doctrine that Christian morality has nothing to do with politics is vigorously upheld by the Nazi régime. This view is basically different from that of the realist who makes morality a function of politics. But in the field of politics it tends to become indistinguishable from realism.

The theory of the divorce between the spheres of politics and morality is superficially attractive, if only because it evades the insoluble problem of finding a moral justification for the use of force.[2] But it is not ultimately satisfying. Both non-resistance and anarchism are counsels of despair, which appear to find widespread acceptance only where men feel hopeless of achieving anything by political action ; and the attempt to keep God and Caesar in watertight compartments runs too much athwart the deep-seated desire of the human mind to reduce its view of the world to some kind of moral order. We are not in the long run satisfied to believe that what is politically good is morally bad ; [3] and since we can neither moralise power nor expel power from politics, we are faced with a dilemma which cannot be completely resolved. The planes of utopia and of reality never coincide. The ideal cannot be institutional-ised, nor the institution idealised. " Politics ", writes Dr. Niebuhr, " will, to the end of history, be an area where con-science and power meet, where the ethical and coercive factors of human life will interpenetrate and work out their tentative and uneasy compromises." [4] The compromises, like solutions of other human problems, will remain uneasy and tentative.

[1] Bernhardi, *Germany and the Next War* (Engl. transl.), p. 29.

[2] " Force in the right place ", as Mr. Maxton once said in the House of Commons. is a meaningless conception, " because the right place for me is exactly where I want to use it, and for him also, and for everyone else " (House of Commons, November 7, 1933 : *Official Record*, col. 130). Force in politics is always the instrument of some kind of group interest.

[3] Acton was fond of saying that " great men are almost always bad men ", and quotes Walpole's dictum that " no great country was ever saved by good men " (*History of Freedom*, p. 219). Rosebery shewed more acuteness when he remarked that " there is one question which English people ask about great men : Was he 'a good man ' ? " (*Napoleon : The Last Phase*, p. 364).

[4] R. Niebuhr, *Moral Man and Immoral Society*, p. 4.

But it is an essential part of any compromise that both factors shall be taken into account.

We have now therefore to analyse the part played in international politics by these two cardinal factors : power and morality

CHAPTER 8

POWER IN INTERNATIONAL POLITICS

POLITICS are, then, in one sense always power politics. Common usage applies the term " political " not to all activities of the state, but to issues involving a conflict of power. Once this conflict has been resolved, the issue ceases to be " political " and becomes a matter of administrative routine. Nor is all business transacted between states " political ". When states co-operate with one another to maintain postal or transport services, or to prevent the spread of epidemics or suppress the traffic in drugs, these activities are described as " non-political " or " technical ". But as soon as an issue arises which involves, or is thought to involve, the power of one state in relation to another, the matter at once becomes " political ". While politics cannot be satisfactorily defined exclusively in terms of power, it is safe to say that power is always an essential element of politics. In order to understand a political issue, it is not enough (as it would be in the case of a technical or a legal issue) to know what the point at issue is. It is necessary also to know between whom it has arisen. An issue raised by a small number of isolated individuals is not the same political fact as the same issue raised by a powerful and well-organised trade union. A political issue arising between Great Britain and Japan is something quite different from what may be formally the same issue between Great Britain and Nicaragua. " Politics begin where the masses are," said Lenin, " not where there are thousands, but where there are millions, that is where serious politics begin." [1]

There have been periods of history when it might have been superfluous to dwell on this obvious fact, and when Engels' dictum that " without force and iron ruthlessness nothing is achieved in history " [2] would have passed as a platitude. But in the comparatively well-ordered world of nineteenth-century liberalism, subtler forms of compulsion successfully concealed from the unsophisticated the continuous but silent workings of

[1] Lenin, *Selected Works* (Engl. transl.), vii. p. 295.
[2] Marx and Engels, *Works* (Russian ed.), vii. p. 212.

political power ; and in democracies, at any rate, this conceal-
ment is still partially effective.[1] After the first world war, the
liberal tradition was carried into international politics. Utopian
writers from the English-speaking countries seriously believed
that the establishment of the League of Nations meant the
elimination of power from international relations, and the substitution of discussion for armies and navies. " Power politics "
were regarded as a mark of the bad old times, and became a
term of abuse. That this belief should have persisted for more
than ten years was due to the circumstance that the Great
Powers whose main interest was the preservation of the *status
quo* enjoyed throughout that time a virtual monopoly of power.
A game of chess between a world-champion and a schoolboy
would be so rapidly and so effortlessly won that the innocent
onlooker might be pardoned for assuming that little skill was
necessary to play chess. In the same way, the simple-minded
spectator of the game of international politics could assume,
between 1920 and 1931, that power played little part in the
game. What was commonly called the " return to power
politics " in 1931 was, in fact, the termination of the monopoly
of power enjoyed by the *status quo* Powers. Stalin's lament
that " *in our days* it is not the custom to reckon with the
weak ", and Neville Chamberlain's remark that " *in the world as
we find it to-day* an unarmed nation has little chance of making
its voice heard ",[2] were curious tributes — more surprising in
the professed Marxist than in the inheritor of a British nineteenth-century tradition — to the illusion that there was once
a time when weak and unarmed countries played an effective
role in international politics.

   The assumption of the elimination of power from politics

[1] Even Lord Baldwin committed himself in 1925 to the dangerous half-truth
that " democracy is government by discussion, by talk " (*On England*, p. 95).
In a recent letter to *The Times*, Mr. Frederic Harrison remarks of the British
Commonwealth of Nations that it " is not founded on conquest and held together
by force of arms. It has been acquired not by the force of our navy and our army
but by force of character, and knit together by ties of sympathy, of a common
interest, a common language and a common history " (*The Times*, June 30, 1938).
This, too, is a dangerous half-truth, which burkes the other and equally important
half of the truth, i.e. that the British Commonwealth is held together by the immense
military and economic power of Great Britain and would at once dissolve if that
power were lost.

[2] Report to the Sixteenth Congress of the Russian Communist Party reprinted
in *L'Union Soviétique et la Cause de la Paix*, p. 25 ; *The Times*, June 26, 1939.
The italics have been inserted in both cases.

could only result from a wholly uncritical attitude towards political problems. In the affairs of the League of Nations, formal equality and the participation of all in debate did not render the power factor any less decisive. The founders of the League themselves entertained no such illusion. House originally thought that only Great Powers should be admitted to the League at all.[1] In the earliest British and American drafts of the Covenant, it was contemplated that membership of the Council of the League would be limited to Great Powers ; and Lord Cecil noted on one of these drafts that " the smaller Powers would in any case not exercise any considerable influence ".[2] This prevision was fulfilled. An Italian delegate testified that during the long period of his regular attendances at Geneva he " never saw a dispute of any importance settled otherwise than by an agreement between the Great Powers ", and that the procedure of the League was " a system of detours, all of which lead to one or other of these two issues : agreement or disagreement between Great Britain, Italy, France and Germany ".[3] " Despite our juridical equality here," said Mr. De Valera a little later, " in matters such as European peace the small states are powerless." [4] The decisions on the application of sanctions against Italy in the winter of 1935–36 were, in effect, taken solely by Great Britain and France, the possessors of effective military and economic power in the Mediterranean. The minor Powers followed their lead ; and one of them was actually " compensated " by Great Britain and France for so doing.

Nor was it only at Geneva that the weak Powers set their course to match that of the strong. When Great Britain took her currency off the gold standard in September 1931, several minor Powers were obliged to follow her example. When France abandoned the gold standard in September 1936, Switzerland and Holland — the last free gold countries — were compelled to follow suit, and several other smaller countries had to alter the value of their currencies. When France was militarily supreme in Europe in the nineteen-twenties, a number

[1] *Intimate Papers of Colonel House*, ed. C. Seymour, iv. p. 24.

[2] Miller, *The Drafting of the Covenant*, ii. p. 61. The result of the subsequent enlargements of the Council has already been mentioned (p. 29 note 1).

[3] *The Foreign Policy of the Powers* (1935 : reprinted from *Foreign Affairs*), pp. 86-7.

[4] *League of Nations : Sixteenth Assembly*, Part II. p. 49.

of smaller Powers grouped themselves under her aegis. When German military strength eclipsed that of France, most of these Powers made declarations of neutrality or veered to the side of Germany. The alleged " dictatorship of the Great Powers ", which is sometimes denounced by utopian writers as if it were a wicked policy deliberately adopted by certain states, is a fact which constitutes something like a " law of nature " in international politics.

It is necessary at this point to dispel the current illusion that the policy of those states which are, broadly speaking, satisfied with the *status quo* and whose watchword is " security ", is somehow less concerned with power than the policy of the dissatisfied states, and that the popular phrase " power politics " applies to the acts of the latter but not to those of the former. This illusion, which has an almost irresistible attraction for the publicists of the satisfied Powers, is responsible for much confused thinking about international politics. The pursuit of " security " by satisfied Powers has often been the motive of flagrant examples of power politics. In order to secure themselves against the revenge of a defeated enemy, victorious Powers have in the past resorted to such measures as the taking of hostages, the mutilation or enslavement of males of military age or, in modern times, the dismemberment and occupation of territory or forced disarmament. It is profoundly misleading to represent the struggle between satisfied and dissatisfied Powers as a struggle between morality on one side and power on the other. It is a clash in which, whatever the moral issue, power politics are equally predominant on both sides.

The history of the Locarno Treaty is a simple and revealing illustration of the working of power politics. The first proposal for a treaty guaranteeing Germany's western frontier was made by Germany in December 1922, and was emphatically rejected by Poincaré. At this period (it was the eve of the Ruhr invasion), Germany had everything to fear from France, and France nothing to fear from a helpless Germany ; and the treaty had no attraction for France. Two years later the position had changed. The Ruhr invasion had brought little profit to France, and had left her perplexed as to the next step. Germany might one day be powerful again. Germany, on the other hand, still feared the military supremacy of France, and hankered after a guarantee. It was the psychological moment when French fear of Germany was about equally balanced by Germany's fear of France ; and a treaty which had not been possible two years before, and would not have been possible five years later, was now welcome to both. Moreover, the power interests of Great Britain coincided with those of Germany. Germany had abandoned hope of securing a revision of her western, but not of her other, frontiers. Great Britain was prepared to guarantee Germany's western, but not her other, frontiers. Germany, anxious to expedite the withdrawal of the Allied army from the Rhineland, had as yet no hope of breaking down the restrictions imposed by the demilitarisation clauses of the Versailles Treaty ; and she was therefore quite prepared to purchase the new agreement by reaffirming her acceptance of those clauses and placing them under a guarantee.

Such was the background of the famous Locarno Treaty. Its success was a striking one. For years afterwards, attempts were made to repeat it in other fields. Mediterranean and Eastern European " Locarnos " were canvassed ; and their failure to materialise disappointed and puzzled people who believed that international problems everywhere could be solved by devices of the same standard pattern, and who failed to understand that the Locarno Treaty was an expression of the power politics of a particular period and locality. Ten years after its conclusion, the delicate balance on which it rested had disappeared. France feared Germany more than ever. But Germany no longer feared anything from France. The treaty no longer had any meaning for Germany save as an affirmation of the demilitarisation clauses of the Versailles Treaty which she could now hope to overthrow. The only part of the Locarno Treaty which still corresponded to the situation of power politics was the British guarantee to France and Belgium. This was repeated by Great Britain after the rest of the treaty had been denounced by Germany. The history of Locarno is a classic instance of power politics. It remains incomprehensible to those who seek uniform *a priori* solutions of the problem of security, and regard power politics as an abnormal phenomenon visible only in periods of crisis.

Failure to recognise that power is an essential element of politics has hitherto vitiated all attempts to establish international forms of government, and confused nearly every attempt to discuss the subject. Power is an indispensable instrument of government. To internationalise government in any real sense means to internationalise power ; and international government is, in effect, government by that state which supplies the power necessary for the purpose of governing. The international governments set up by the Versailles Treaty in various parts of Europe were temporary in character, and had not therefore to face the problems of a long-term policy. But even these illustrate the intimate connexion between government and power. The Inter-Allied High Commission, which exercised in the occupied Rhineland such functions of government as were necessary for the security of the Allied troops, worked smoothly so long as British and French policies coincided. When the Ruhr crisis caused a serious difference of opinion between the British and French Governments, French policy was applied in the zones occupied by French and Belgian troops and British policy in the zone occupied by British troops, the policy of the government being determined by the nationality of the power on which it rested. The Inter-Allied Commission appointed to conduct the plebiscite in Upper Silesia pursued the French policy of favouring Poland so long as the Allied troops on which its authority depended were supplied almost exclusively by France. This policy was corrected only when British troops were sent to the area. The effective control of any government depends on the source of its power.

The problem of international government and power was raised in a more acute form by the mandates system and by the proposal frequently put forward that the government of some or all colonial territories shall be " internationalised ". We are here faced by an issue of permanent government, involving the formulation of long-term policy, and different in kind from that of temporary international collaboration between allies under stress of war or for the purpose of implementing a treaty jointly imposed. Its nature may be illustrated from the case of Palestine. Policy in Palestine was dependent on the amount of military force available for use there, and had therefore to be determined not by the Mandates Commission, which had no power at its disposal, but by the British Government, which supplies the power ; for whatever view might be taken by the Mandates Commission, it was unthinkable that British troops could be used to carry out a policy of which the British Government or the British electorate did not approve.[1] Under any international system of government, policy would depend, at critical moments, on the decision of the state supplying the power on which the authority of the government depended. If, as would almost inevitably happen, the control of an international territory were divided geographically among the forces of different states, the different zones would, in periods of international discord, pursue discordant policies : and the old international rivalries would recur in a new and equally dangerous form. Problems of economic development would be not less baffling. The international administration of colonial areas, wrote Lugard, himself an experienced and enlightened administrator, " would paralyse all initiative by the dead hand of a super-bureaucracy devoid of national sentiment and stifling to all patriotism, and would be very disadvantageous to the countries concerned ".[2] Any real international government is impossible so long as power, which is an essential condition of government, is organised nationally. The international secretariat of the League of Nations was able to function precisely because it was a non-political civil service, had no responsibility for policy, and was therefore independent of power.

Political power in the international sphere may be divided, for purposes of discussion, into three categories : (a) military power, (b) economic power, (c) power over opinion. We shall find, however, that these categories are closely interdependent ; and though they are theoretically separable, it is difficult in practice to imagine a country for any length of time possessing one kind of power in isolation from the others. In its essence, power is an indivisible whole. " The laws of social dynamics ", a recent critic has said, " are laws which can only be stated in terms of power, not in terms of this or that form of power." [3]

[1] In 1926, when Palestine was discussed by the Mandates Commission, M. Rappard " thought that the Mandatory would incur grave responsibility if it found itself one day faced with the impossibility of preventing a pogrom owing to insufficient troops. Its responsibility, indeed, would be shared by the Mandates Commission, if that Commission had not pointed out this danger " (*Permanent Mandates Commission, Minutes of Ninth Session*, p. 184). The responsibility of the Commission was thus limited to " pointing out ".

[2] Lugard, *The Dual Mandate in Tropical Africa*, p. 53.

[3] B. Russell, *Power*, p. 11. I owe to this book, which is an able and stimulating analysis of power as " the fundamental concept in social science ", the tripartite classification of power adopted above.

(a) MILITARY POWER

The supreme importance of the military instrument lies in the fact that the *ultima ratio* of power in international relations is war. Every act of the state, in its power aspect, is directed to war, not as a desirable weapon, but as a weapon which it may require in the last resort to use. Clausewitz's famous aphorism that " war is nothing but the continuation of political relations by other means " has been repeated with approval both by Lenin and by the Communist International ; [1] and Hitler meant much the same thing when he said that "an alliance whose object does not include the intention to fight is meaningless and useless ".[2] In the same sense, Mr. Hawtrey defines diplomacy as " potential war ".[3] These are half-truths. But the important thing is to recognise that they are true War lurks in the background of international politics just as revolution lurks in the background of domestic politics. There are few European countries where, at some time during the past thirty years, potential revolution has not been an important factor in politics ; [4] and the international community has in this respect the closest analogy to those states where the possibility of revolution is most frequently and most conspicuously present to the mind.

Potential war being thus a dominant factor in international politics, military strength becomes a recognised standard of political values. Every great civilisation of the past has enjoyed in its day a superiority of military power. The Greek city-state rose to greatness when its hoplite armies proved more than a match for the Persian hordes. In the modern world, Powers (the word itself is significant enough) are graded according to the quality and the supposed efficiency of the military equipment, including man-power, at their disposal. Recognition as a Great Power is normally the reward of fighting a successful large-scale war. Germany after the Franco-Prussian War, the United States after the war with Spain, and Japan after the

---

[1] Lenin, *Collected Works* (Engl. transl.), xviii. p. 97 ; Theses of the Sixth Congress of Comintern quoted in Taracouzio, *The Soviet Union and International Law*, p. 436.

[2] Hitler, *Mein Kampf*, p. 749.

[3] R. G. Hawtrey, *Economic Aspects of Sovereignty*, p. 107.

[4] It is perhaps necessary to recall the part played in British politics in 1914 by the threat of the Conservative Party to support revolutionary action in Ulster.

Russo-Japanese War are familiar recent instances. The faint doubt attaching to Italy's status as a Great Power is partly due to the fact that she has never proved her prowess in a first-class war. Any symptom of military inefficiency or unpreparedness in a Great Power is promptly reflected in its political status. The naval mutiny at Invergordon in September 1931 was the final blow to British prestige which compelled Great Britain to devalue her currency. The execution of the leading Soviet generals for alleged treason in June 1937 was thought to reveal so much weakness in the Soviet military machine that the political influence of Soviet Russia suffered a sudden and severe slump. Statesmen of all the Great Powers periodically make speeches extolling the efficiency of their armies, navies and air forces ; and military parades and reviews are organised in order to impress the world with the military strength and consequent political standing of the nation. In international crises, fleets, troops or air squadrons shew themselves conspicuously at crucial points for the same purpose.

These facts point the moral that foreign policy never can, or never should, be divorced from strategy. The foreign policy of a country is limited not only by its aims, but also by its military strength or, more accurately, by the ratio of its military strength to that of other countries. The most serious problem involved in the democratic control of foreign policy is that no government can afford to divulge full and frank information about its own military strength, or all the knowledge it possesses about the military strength of other countries. Public discussions of foreign policy are therefore conducted in partial or total ignorance of one of the factors which must be decisive in determining it. A constitutional rule of long standing precludes private members of Parliament from proposing motions which entail public expenditure. The same restraint might well be exercised in advocating policies which entail risk of war ; for only the government and its advisers can assess the chances with anything like complete knowledge of the relevant facts. Many contemporary books and speeches about international politics are reminiscent of those ingenious mathematical problems which the student is invited to solve by ignoring the weight of the elephant. The solutions proposed are neat and accurate on the abstract plane, but are obtained by leaving out of account the vital strategic factor. Even so important, and in many ways so valuable, a work as the annual *Survey of International Affairs* frequently soars into the realms of fancy when it embarks on criticism of policy, precisely because it neglects those military limitations which are always present to the minds of those who have to solve problems of foreign policy in real life. If every prospective writer on international affairs in the last twenty years had taken a compulsory course in elementary strategy, reams of nonsense would have remained unwritten.

Military power, being an essential element in the life of the state, becomes not only an instrument, but an end in itself. Few of the important wars of the last hundred years seem to have been waged for the deliberate and conscious purpose of increasing either trade or territory. The most serious wars are fought in order to make one's own country militarily stronger or, more often, to prevent another country from becoming militarily stronger, so that there is much justification for the epigram that "the principal cause of war is war itself".[1] Every stage in the Napoleonic Wars was devised to prepare the way for the next stage : the invasion of Russia was undertaken in order to make Napoleon strong enough to defeat Great Britain. The Crimean War was waged by Great Britain and France in order to prevent Russia from becoming strong enough to attack their Near Eastern possessions and interests at some future time. The origin of the Russo-Japanese War of 1904–5 is described as follows in a note addressed to the League of Nations by the Soviet Government in 1924 : "When the Japanese torpedo-boats attacked the Russian fleet at Port Arthur in 1904, it was clearly an act of aggression from a technical point of view, but, politically speaking, it was an act caused by the aggressive policy of the Tsarist Government towards Japan, who, in order to forestall the danger, struck the first blow at her adversary ".[2] In 1914, Austria sent an ultimatum to Servia because she believed that Servians were planning the downfall of the Dual Monarchy ; Russia feared that Austria-Hungary, if she defeated Servia, would be strong enough to menace her ; Germany feared that Russia, if she defeated Austria-Hungary, would be strong enough to menace her ; France had long believed that Germany, if she defeated

[1] R. G. Hawtrey, *Economic Aspects of Sovereignty*, p. 105.
[2] *League of Nations : Official Journal*, May 1924, p. 578.

Russia, would be strong enough to menace her, and had therefore concluded the Franco-Russian alliance ; and Great Britain feared that Germany, if she defeated France and occupied Belgium, would be strong enough to menace her. Finally, the United States came to fear that Germany, if she won the war, would be strong enough to menace them. Thus the war, in the minds of all the principal combatants, had a defensive or preventive character. They fought in order that they might not find themselves in a more unfavourable position in some future war. Even colonial acquisitions have often been prompted by the same motive. The consolidation and formal annexation of the original British settlements in Australia were inspired by fear of Napoleon's alleged design to establish French colonies there. Military, rather than economic, reasons dictated the capture of German colonies during the war of 1914 and afterwards precluded their return to Germany.

It is perhaps for this reason that the exercise of power always appears to beget the appetite for more power. There is, as Dr. Niebuhr says, " no possibility of drawing a sharp line between the will-to-live and the will-to-power ".[1] Nationalism, having attained its first objective in the form of national unity and independence, develops almost automatically into imperialism. International politics amply confirm the aphorisms of Machiavelli that " men never appear to themselves to possess securely what they have unless they acquire something further from another ",[2] and of Hobbes that man " cannot assure the power and means to live well which he hath present, without the acquisition of more ".[3] Wars, begun for motives of security, quickly become wars of aggression and self-seeking. President McKinley invited the United States to intervene in Cuba against Spain in order " to secure a full and final termination of hostilities between the Government of Spain and the people of Cuba and to secure on the island the establishment of a stable government ".[4] But by the time the war was over the temptation to self-aggrandisement by the annexation of the Philippines had become irresistible. Nearly every country participating in the first world war regarded it initially as a

---

[1] R. Niebuhr, *Moral Man and Immoral Society*, p. 42.
[2] Machiavelli, *Discorsi*, 1. i. ch. v.
[3] Hobbes, *Leviathan*, ch. xi.
[4] *British and Foreign State Papers*, ed. Hertslet, xc. p. 811.

war of self-defence ; and this belief was particularly strong
on the Allied side. Yet during the course of the war, every
Allied Government in Europe announced war aims which
included the acquisition of territory from the enemy Powers.
In modern conditions, wars of limited objective have become
almost as impossible as wars of limited liability. It is one
of the fallacies of the theory of collective security that war
can be waged for the specific and disinterested purpose of
" resisting aggression ". Had the League of Nations in the
autumn of 1935, under the leadership of Great Britain, em-
barked on " military sanctions " against Italy, it would have
been impossible to restrict the campaign to the expulsion of
Italian troops from Abyssinia. Operations would in all probability have led to the occupation of Italy's East African
colonies by Great Britain and France, of Trieste, Fiume and
Albania by Yugoslavia, and of the islands of the Dodecanese
by Greece or Turkey or both ; and war aims would have been
announced, precluding on various specious grounds the
restoration of these territories to Italy. Territorial ambitions
are just as likely to be the product as the cause of war.

(b) ECONOMIC POWER

Economic strength has always been an instrument of political
power, if only through its association with the military instrument. Only the most primitive kinds of warfare are altogether
independent of the economic factor. The wealthiest prince or
the wealthiest city-state could hire the largest and most efficient
army of mercenaries ; and every government was therefore
compelled to pursue a policy designed to further the acquisition
of wealth. The whole progress of civilisation has been so
closely bound up with economic development that we are not
surprised to trace, throughout modern history, an increasingly
intimate association between military and economic power. In
the prolonged conflicts which marked the close of the Middle
Ages in Western Europe, the merchants of the towns, relying
on organised economic power, defeated the feudal barons, who
put their trust in individual military prowess. The rise of
modern nations has everywhere been marked by the emergence
of a new middle class economically based on industry and trade.
Trade and finance were the foundation of the short-lived

political supremacy of the Italian cities of the Renaissance and later of the Dutch. The principal international wars of the period from the Renaissance to the middle of the eighteenth century were trade wars (some of them were actually so named). Throughout this period, it was universally held that, since wealth is a source of political power, the state should seek actively to promote the acquisition of wealth; and it was believed that the right way to make a country powerful was to stimulate production at home, to buy as little as possible from abroad, and to accumulate wealth in the convenient form of precious metals. Those who argued in this way afterwards came to be known as mercantilists. Mercantilism was a system of economic policy based on the hitherto unquestioned assumption that to promote the acquisition of wealth was part of the normal function of the state.

The Separation of Economics from Politics

The *laissez-faire* doctrine of the classical economists made a frontal attack on this assumption. The principal implications of *laissez-faire* have already been discussed. Its significance in the present context is that it brought about a complete theoretical divorce between economics and politics. The classical economists conceived a natural economic order with laws of its own, independent of politics and functioning to the greatest profit of all concerned when political authority interfered least in its automatic operation. This doctrine dominated the economic thought, and to some extent the economic practice (though far more in Great Britain than elsewhere), of the nineteenth century. The theory of the nineteenth century liberal state presupposed the existence side by side of two separate systems. The political system, which was the sphere of government, was concerned with the maintenance of law and order and the provision of certain essential services, and was thought of mainly as a necessary evil. The economic system, which was the preserve of private enterprise, catered for the material wants and, in doing so, organised the everyday lives of the great mass of the citizens.[1] In current English theory,

[1] The distinction between the two systems is implicit in Saint-Simon's prediction that the "industrial régime" will succeed the "military régime", and "administration" replace "government", better known in the form given to it by Engels

the doctrine of the separation of politics and economics was sometimes carried to astonishing lengths. " Is it true ", asked Sir Norman Angell shortly before the first world war, " that wealth and prosperity and well-being depend on the political power of nations, or indeed that one has anything whatever to do with the other ? " [1] And the whole argument depends on the confident assumption that every intelligent reader will answer in the negative. As late as 1915, an English philosopher detected " an ineradicable tendency that, as wealth and its control and enjoyment go to the productive class, so power and prestige go to the professional class ", and regarded this separation of economic from political power as not only ineradicable but " essential to a decent society ".[2]

Even before 1900, a more penetrating analysis might have shewn that the illusion of a divorce between politics and economics was fast breaking down. It is still open to debate whether late nineteenth-century imperialism should be regarded as an economic movement using political weapons, or as a political movement using economic weapons. But that economics and politics marched hand in hand towards the same objective is clear enough. " Is it not precisely the hallmark of British statesmanship ", asked Hitler, " to draw economic advantages from political strength, and to transform every economic gain back into political power ? " [3] The first world war, by overtly reuniting economics and politics, in both domestic and foreign policy, hastened a development which was already on the way. It was now revealed that the nineteenth century, while purporting to take economics altogether out of the political sphere, had in fact forged economic weapons of unparalleled strength for use in the interests of national policy. A German staff officer had remarked to Engels in the 1880's that " the basis of warfare is primarily the general economic life of peoples " ; [4] and this diagnosis was amply confirmed by the experiences of 1914–18. In no previous war had the economic life of belligerent nations been so completely and ruthlessly organised by the political authority. In the age-long alliance

that the " administration of things " will replace the " government of men " (quotations in Halévy, *Ère des Tyrannies*, p. 224).

[1] Angell, *The Great Illusion*, ch. ii.
[2] B. Bosanquet, *Social and International Ideals*, pp. 234-5.
[3] Hitler, *Mein Kampf*, p. 158.
[4] Engels, *Anti-Dühring* (Engl. transl.), p. 195.

between the military and the economic arm, the economic arm for the first time was an equal, if not a superior, partner. To cripple the economic system of an enemy Power was as much a war aim as to defeat his armies and fleets. " Planned economy ", which means the control by the state for political purposes of the economic life of the nation, was a development of the first world war.[1] " War potential " has become another name for economic power.

We have now therefore returned, after the important, but abnormal, *laissez-faire* interlude of the nineteenth century, to the position where economics can be frankly recognised as a part of politics. We can thus resolve the controversy, which is in large part a product of nineteenth-century ideas and terminology, about the so-called economic interpretation of history. Marx was overwhelmingly right when he insisted on the increasing importance of the role played by economic forces in politics ; and since Marx, history can never be written again exactly as it was written before him. But Marx believed, just as firmly as did the *laissez-faire* liberal, in an economic system with laws of its own working independently of the state, which was its adjunct and its instrument. In writing as if economics and politics were separate domains, one subordinate to the other, Marx was dominated by nineteenth-century presuppositions in much the same way as his more recent opponents who are equally sure that " the primary laws of history are political laws, economic laws are secondary ".[2] Economic forces are in fact political forces. Economics can be treated neither as a minor accessory of history, nor as an independent science in the light of which history can be interpreted. Much confusion would be saved by a general return to the term " political economy ", which was given the new science by Adam Smith himself and not abandoned in favour of the abstract " eco-

---

[1] Planned economy has been developed not only by international frictions, but by social frictions within the state. It can therefore be logically regarded both as a nationalist policy (" economic nationalism ") and as a socialist policy. The second aspect was irrelevant to my present argument, and has therefore been passed over in the text. According to Bruck (*Social and Economic History of Germany*, p. 157), the term *Planwirtschaft* was invented in Germany during the first world war. But the phrase *der staatliche Wirtschaftsplan* occurs in a composite *Grundriss der Sozialökonomik* (i. 424), published at Tübingen just before the war, in the general sense of " state economic policy ".

[2] Moeller van den Bruck, *Germany's Third Empire*, p. 50. The idea is a commonplace of National Socialist and Fascist writers.

nomics ", even in Great Britain itself, till the closing years of the nineteenth century.[1] The science of economics presupposes a given political order, and cannot be profitably studied in isolation from politics.

Some Fallacies of the Separation of Economics from Politics

It would have been unnecessary to dwell at length on this point if its importance had been either purely historical or purely theoretical. The illusion of a separation between politics and economics — a belated legacy of the *laissez-faire* nineteenth century — had ceased to correspond to any aspect of current reality. But it continued to persist in thought about international politics, where it created no little confusion. An immense amount of discussion was devoted to the meaningless question whether (as the Economic Conference of 1927 supposed[2]) our political troubles have economic causes or whether (as the Van Zeeland report suggested[3]) our economic troubles have political causes, and to the equally meaningless conundrum whether the problem of raw materials is political or economic. Similar confusion was produced by the declaration of the British Government in 1922 that the rate of Jewish immigration into Palestine would be determined by " the economic capacity of the country ", supplemented in 1931 by the further statement that " the considerations relevant to the limits of absorptive capacity are purely economic considerations ". It was not until 1937 that a Royal Commission discovered that " since Arabs are hostile to Jewish immigration, the factor of ' hostility between the two peoples ' necessarily assumes immediate economic importance ".[4] Indeed every issue of migration and refugees has been complicated by the supposition that there is some objective economic test of absorptive capacity. The

[1] In Germany, " political economy " was at first translated *Nationalökonomie*, which was tentatively replaced in the present century by *Sozialökonomie*.

[2] " Economic conflicts and divergence of economic interest are perhaps the most serious and most permanent of all the dangers which are likely to threaten the peace of the world " (*League of Nations* : C.E.I. 44, p. 7).

[3] " I have deliberately debarred myself from touching on the strictly political aspects. . . . It is, however, impossible to ignore the fact that we are working in their shadow " (*Report . . . on the Possibility of Obtaining a General Reduction of the Obstacles to International Trade*, Cmd. 5648).

[4] All these passages are quoted from the Report of the Palestine Royal Commission of 1937, Cmd. 5479, pp. 298-300.

conflict between two opposite and equally defensible interpretations of the promise in the Treaty of Neuilly " to ensure the economic outlets of Bulgaria to the Aegean Sea " was another instance of confusion arising from the too light-hearted use of this elusive word. Attempts to solve international problems by the application of economic principles divorced from politics are doomed to sterility.

The most conspicuous practical failure caused by the persistence of this nineteenth-century illusion was the breakdown of League sanctions in 1936. Careful reading of the text of Article 16 of the Covenant acquits its framers of responsibility for the mistake. Paragraph 1 prescribes the economic weapons, paragraph 2 the military weapons, to be employed against the violator of the Covenant. Paragraph 2 is clearly complementary to paragraph 1, and assumes as a matter of course that, in the event of an application of sanctions, " armed forces " would be required " to protect the Covenants of the League ". The only difference between the two paragraphs is that, whereas all members of the League would have to apply the economic weapons, it would be natural to draw the necessary armed forces from those members which possessed them in sufficient strength and in reasonable geographical proximity to the offender.[1] Subsequent commentators, obsessed with the assumption that economics and politics were separate and separable things, evolved the doctrine that paragraphs 1 and 2 of Article 16 were not complementary, but alternative, the difference being that " economic sanctions " were obligatory and " military sanctions " optional. This doctrine was eagerly seized on by the many who felt that the League might conceivably be worth a few million pounds worth of trade, but not a few million human lives ; and in the famous 1934 Peace Ballot in Great Britain, some two million deluded voters expressed simultaneously their approval of economic, and their disapproval of military, sanctions. " One of the many conclusions to which I have been drawn ", said Lord Baldwin at this time, " is that

[1] This interpretation is confirmed by the report of the Phillimore Committee, on whose proposals the text of Article 16 was based. The Committee " considered financial and economic sanctions as being simply the contribution to the work of preventing aggression which might properly be made by countries which were not in a position to furnish actual military aid " (*International Sanctions : Report by a Group of Members of the Royal Institute of International Affairs*, p. 115. where the relevant texts are examined)

there is no such thing as a sanction which will work, which does not mean war." [1]   But the bitter lesson of 1935–36 was needed to drive home the truth that in sanctions, as in war, the only motto is " all or nothing ", and that economic power is impotent if the military weapon is not held in readiness to support it.[2] Power is indivisible ; and the military and economic weapons are merely different instruments of power.[3]

A different, and equally serious, form in which this illusory separation of politics and economics can be traced is the popular phraseology which distinguishes between " power " and " welfare ", between " guns " and " butter ".   " Welfare arguments are ' economic ' ", remarks an American writer, " power arguments are ' political '." [4]   This fallacy is particularly difficult to expose because it appears to be deducible from a familiar fact.   Every modern government and every parliament is continually faced with the dilemma of spending money on armaments or social services ; and this encourages the illusion that the choice really lies between " power " and " welfare ", between political guns and economic butter.   Reflexion shews, however, that this is not the case.   The question asked never takes the form, Do you prefer guns or butter ?   For everyone (except a handful of pacifists in those Anglo-Saxon countries which have inherited a long tradition of uncontested security) agrees that, in case of need, guns must come before butter The question asked is always either, Have we already sufficient guns to enable us to afford some butter ?   or, Granted that we need x guns, can we increase revenue sufficiently to afford more butter as well ?   But the neatest exposure of this fallacy

---

[1] House of Commons, May 18, 1934 : *Official Report*, col. 2139.

[2] It is not, of course, suggested that the military weapon must always be used. The British Grand Fleet was little used in the first world war.   But it would be rash to assume that the result would have been much the same if the British Government had not been prepared to use it.   What paralysed sanctions in 1935–36 was the common knowledge that the League Powers were not prepared to use the military weapon.

[3] It is worth noting that Stresemann was fully alive to this point when Germany entered the League of Nations.   When the Secretary-General argued that Germany, if she contracted out of military sanctions, could still participate in economic sanctions, Stresemann replied : " We cannot do that either ; if we take part in an economic boycott of a powerful neighbour, a declaration of war against Germany might be the consequence, since the exclusion of another country from intercourse with a nation of sixty million citizens would be a hostile act " (*Stresemann's Diaries and Papers* (Engl. transl.), ii. p. 69).

[4] F. L. Schuman, *International Politics*, p. 356.

comes from the pen of Professor Zimmern ; and the exposure is none the less effective for being unconscious. Having divided existing states on popular lines into those which pursue " welfare " and those which pursue " power ", Professor Zimmern revealingly adds that " the welfare states, taken together, enjoy a preponderance of power and resources over the power states ",[1] thereby leading us infallibly to the correct conclusion that " welfare states " are states which, already enjoying a preponderance of power, are not primarily concerned to increase it, and can therefore afford butter, and " power states " those which, being inferior in power, are primarily concerned to increase it, and devote the major part of their resources to this end. In this popular terminology, " welfare states " are those which possess preponderant power, and " power states " those which do not. Nor is this classification as illogical as it may seem. Every Great Power takes the view that the minimum number of guns necessary to assert the degree of power which it considers requisite takes precedence over butter, and that it can only pursue " welfare " when this minimum has been achieved. For many years prior to 1933, Great Britain, being satisfied with her power, was a " welfare state " After 1935, feeling her power contested and inadequate, she became a " power state " ; and even the Opposition ceased to press with any insistence the prior claim of the social services. The contrast is not one between " power " and " welfare ", and still less between " politics " and " economics ", but between different degrees of power. In the pursuit of power, military and economic instruments will both be used.

Autarky

Having thus established that economics must properly be regarded as an aspect of politics, we may divide into two broad categories the methods by which economic power is pressed into the service of national policy. The first will contain those measures whose purpose is defined by the convenient word autarky ; the second, economic measures directly designed to strengthen the national influence over other countries

Autarky, or self-sufficiency, was one of the aims of the mercantilist policy, and has indeed been pursued by states from

[1] Zimmern, *Quo Vadimus ?* p. 41.

the earliest times. But the problem of autarky is nevertheless distinctively modern. In the Middle Ages, autarky was a natural and necessary condition of economic life ; for the long-distance transport of any goods other than those of small bulk and great value was unremunerative. From the close of the Middle Ages, transport gradually became safer, cheaper and more rapid. Countries became less completely self-dependent ; and a rising standard of life was based in part on the international exchange of specialised products. But it is only within the last hundred years that the coming of steam has made transport by land and sea so rapid and cheap that the cost of transport of most commodities is now insignificant in relation to the cost of production, and it is in many cases immaterial whether an article is produced 500 or 5000 miles from the point where it will be used or consumed. Mass-production methods, under which commodities become cheaper the more of them are produced in the same place, have further promoted concentration. Not only are our needs to-day more highly specialised than ever before, but we live in a world where, for the first time in history, it might, from the standpoint of cost, be possible — and perhaps even desirable — to grow all the wheat consumed by the human race in Canada, and all the wool in Australia, to manufacture all the motor cars in Detroit and all the cotton clothing in England or Japan. Internationally, the consequences of absolute *laissez-faire* are as fantastic and as unacceptable as are the consequences of *laissez-faire* within the state. In modern conditions the artificial promotion of some degree of autarky is a necessary condition of orderly social existence

Autarky is, however, not only a social necessity, but an instrument of political power. It is primarily a form of preparedness for war. In the mercantilist period, it was commonly asserted, both in Britain and elsewhere, that the military power of the state depended on the production of manufactured goods. Adam Smith made his famous exceptions to the doctrine of *laissez-faire* when he approved of the British Navigation Act and the bounties on British sail-cloth and British gunpowder. But the principle of autarky received its classic definition from the pen of Alexander Hamilton, who in 1791, being then Secretary of the United States Treasury, made a report to the House of Representatives which enunciates, in words which might have been written to-day, the whole modern doctrine of autarky. Hamilton had been instructed to advise on " the means of promoting such [manufactures] as will tend to render the United States independent of foreign nations for military and other essential supplies ". One short passage may be quoted from the report :

> Not only the wealth but the independence and security of a country appear to be materially connected with the prosperity of manufactures. Every nation, with a view to these great objects, ought to endeavour to possess within itself all the essentials of national supply. . . . The extreme embarrassments of the United States during the late war, from an incapacity of supplying themselves, are still a matter of keen recollection ; a future war might be expected to exemplify the mischief and dangers of a situation to which that capacity is still, in too great a degree, applicable, unless changed by timely and vigorous action.

And Hamilton went on to examine in turn all the methods by which the desired result might be attained — duties, prohibitions, bounties and premiums.[1] In Germany, just fifty years later, List argued that " on the development of the German protective system depend the existence, the independence and the future of the German nationality " ;[2] and in the latter half of the nineteenth century successive Prussian victories drove home the intimate connexion between a highly developed industrial system and military power.

Throughout this period Great Britain, in virtue of her industrial supremacy, enjoyed virtually complete autarky in all industrial products, though not in the raw materials required to produce them. In food supplies, she ceased to be self-supporting about 1830. But this defect was in large part remedied by her naval power, the maintenance of which became one of her chief preoccupations. A Royal Commission on the Supply of Food and Raw Materials in Time of War, which reported in 1905, discussed, but rejected, plans for the precautionary storage in Great Britain of reserve supplies, and did not even discuss any plan for encouraging home production. Complete reliance was placed on the capacity of the navy to

---

[1] *Works of Alexander Hamilton*, iv. pp. 69 *sqq.*
[2] List, *The National System of Political Economy* (Engl. transl.), p. 425.

protect the ordinary channels of trade, and thereby make up for the inevitable absence of sufficient supplies at home [1] The now current view that nineteenth-century statesmen were not alive to the political desirability of autarky, or of some adequate substitute for it, is not borne out by facts.

The effect of the first world war on the whole concept of economics has already been discussed. The impulse which it gave to the pursuit of autarky was immediate and powerful. Blockade, and the diversion of a large part of the world's shipping to the transport of troops and munitions, imposed more or less stringent measures of autarky on both belligerents and neutrals. For four years, the Central Powers were compelled to depend exclusively on their own resources, and to realise in spite of themselves Fichte's ideal of *The Closed Commercial State*. Even for the Allied Powers, the new weapon of the submarine made reliance on overseas imports as an alternative to autarky more precarious than it had hitherto been supposed. Nor did the Allied Governments, at any rate, appear to regard autarky as a regrettable and temporary expedient. In June 1916, they met in Paris to discuss post-war economic policy, and decided " to take the necessary steps without delay to render themselves independent of the enemy countries in so far as regards the raw materials and manu-factured articles essential to their normal economic activities ".[2] In the following year, a British Royal Commission drew up a list of articles in respect of which it had been established " that the possibility of economic pressure from foreign countries controlling supplies of raw materials requires especially to be guarded against, and that government action is most needed in order to promote economic independence " ; and this policy was carried into effect in the Safeguarding of Industries Act of 1921. Where home supplies were not available, the unfettered control of overseas supplies became a primary objective. The desire to control adequate supplies of oil inspired an active British policy in more than one oil-producing country.

Internationally, the important part played by the blockade in winning the war made inevitable the prominence of " economic sanctions " in the constitution of the League of Nations.

---

[1] *Supply of Food and Raw Materials in Time of War*, Cmd. 2644.
[2] The resolutions are printed in *History of the Peace Conference*, ed. Temperley, v. pp. 368-9.

It was clear that blockade was likely to be applied more vigorously than ever in another war; and autarky was developed as the natural defensive armament against the weapon of blockade. The actual use of this weapon against Italy in 1935 added point to the moral. " November 18, 1935, marks the starting point of a new chapter in Italian history ", said Mussolini to the National Guild Assembly on March 23, 1936. " . . . The new phase of Italian history will be determined by this postulate : to secure within the briefest time possible the greatest possible measure of economic independence." There was, in fact, little novelty in this doctrine, which was merely a paraphrase of what had been said by Hamilton, by List, and by the British Royal Commission of 1917. But the growing international tension threw the problem into sharp relief. A well-known American publicist urged the joint buying by Great Britain and the United States of " metals of strategic importance " with the object of " removing the great bulk of these important metals from the markets in which the dictatorial and ' have-not ' Powers must buy them ".[1] " No measure ", added a British writer, " would do more to weaken a German rearmament programme than a British decision to purchase the entire available output of Swedish ore." [2] It scarcely required such warnings to convince governments of the military value of autarky. The development of synthetic materials by Germany and the accumulation by Great Britain of stocks of foodstuffs and essential raw materials were two of many significant symptoms. Autarky, like other elements of power, is expensive. It may cost a country as much to make itself self-supporting in some important commodity as to build a battleship. The expenditure may turn out to be wasteful, and the acquisition not worth the cost. But to deny that autarky is an element of power, and as such desirable, is to obscure the issue.

Economic Power as an Instrument of Policy

The second use of the economic weapon as an instrument of national policy, i.e. its use to acquire power and influence abroad, has been so fully recognised and freely discussed that the briefest summary will suffice here. It takes two principal

[1] W. Y. Elliott in *Political Quarterly*, April–June 1938, p. 181.
[2] G. D. H. Cole in *Political Quarterly*, January–March 1939, p. 65.

forms : (*a*) the export of capital, and (*b*) the control of foreign markets.

(*a*) The export of capital has in recent times been a familiar practice of powerful states. The political supremacy of Great Britain throughout the nineteenth century was closely associated with London's position as the financial centre of the world. Only in Europe, where Great Britain did not aspire to political influence, were British investments insignificant, amounting to not more than 5 per cent of all British capital invested abroad. The rise of the United States to political power in the present century was largely due to their appearance in the market as a large-scale lender, first of all, to Latin America, and since 1914, to Europe. The attainment of political objectives by direct government investment occurred in such cases as the purchase by the British Government of shares in the Suez Canal Company and the Anglo-Iranian Oil Company, or the construction of the Chinese Eastern Railway with Russian Government capital. More often, governments used their power to stimulate investments by banks and private individuals in the interests of national policy. Thus the Franco-Russian alliance was cemented by some £400,000,000 of French capital lent by French investors to the Russian Government. In Germany, " the joint-stock bank was not merely a credit organisation, but a politico-economic instrument ; it was an instrument of Germany's power policy ".[1] The whole policy of nineteenth-century imperialism was based on the development of the backward parts of the world through investment of European capital. Political interests were furthered by private investors enjoying, like the chartered companies of the nineteenth century, government patronage or, more commonly, diplomatic support.[2] Marx described the policy as one of replacing " the feudal method of waging war . . . by the mercantile method, cannons by capital " ; [3] and a new and expressive phrase was coined to describe the " dollar diplomacy " of the United States.

---

[1] W. F. Bruck, *Social and Economic History of Germany*, p. 80.
[2] The whole subject is thoroughly investigated, and innumerable examples are cited, in Eugene Staley, *War and the Private Investor*. Mr. Staley's main conclusion is that official policy has rarely been influenced in an important degree by private investment, but that private investment has again and again been officially directed and encouraged as an instrument of policy.
[3] Marx, *Gesammelte Schriften*, i. p. 84.

The diplomacy of the present administration [said Taft in 1912] has sought to respond to the modern ideas of commercial intercourse. This policy has been characterised as substituting dollars for bullets. It is one that appeals alike to idealistic humanitarian sentiments, to the dictates of sound policy and strategy, and to legitimate commercial aims.[1]

The frequent appearances of the American fleet in Latin American waters (like those of the British fleet elsewhere) shewed, moreover, that, if dollars were a humanitarian substitute for bullets, they could and would be reinforced by bullets in case of political need.

The diminished use after 1919 of capital investment abroad as an instrument of policy was explained by the rapid falling-off in the accumulation of surplus capital throughout the world and the insolvency of many potential borrowers. But numerous familiar examples may still be cited. France strengthened her influence over Poland and the Little Entente by abundant loans and credits, public and private, to these countries. Several governments granted or guaranteed loans to Austria for the political purpose of maintaining Austria's independence ; and in 1931 French financial pressure obliged Austria to abandon the project of a customs union between Austria and Germany. The rapid decline of French influence in Central Europe after 1931 was closely connected with the fact that France, since the crisis, was unable to continue her policy of financial assistance to these countries. When in December 1938 it was announced that the French Schneider-Creusot group had sold its interest in the Skoda works to a Czecho-Slovak group representing the Czecho-Slovak Government, a correspondent of *The Times* commented that " this transaction is another indication of France's retreat from Central Europe, and puts an end to a chapter of French political expansion ".[2] After 1932, when an unofficial embargo was placed on the issue of foreign loans in the British market, it could fairly be said that Great Britain's foreign lending was subject to political supervision. The years 1938 and 1939 saw the grant to Turkey by Great Britain and Germany, and to China by the United States and Great Britain, of " commercial " credits whose political motive was scarcely disguised.

(*b*) The struggle to control foreign markets provides a further illustration of the interaction of politics and economics ; for it is often impossible to decide whether political power is being used to acquire markets for the sake of their economic value, or whether markets are being sought in order to establish and strengthen political power. The struggle for markets has been the most characteristic feature of the economic warfare of the period between the two world wars. It would be wrong to attribute exclusively to political rivalries the intensified pressure to export which manifested itself everywhere. Under the modern structure of industry, the most economical scale of production of many commodities exceeds the consumption capacity of most national markets ; and to sell dear in a protected home market and cheap in a free foreign market (which is the essence of " dumping ") may be perfectly sound policy from the purely commercial standpoint. Yet the use of dumping as an instrument of policy is incontestable ; and powerful countries found their " natural " markets in areas where their political interests lay and where their political influence could be most readily asserted. The principal reason why Central and South-Eastern Europe were Germany's " natural " markets was their accessibility to Germany's military power. German rearmament and German economic penetration of these areas proceeded simultaneously. This was, however, not a new phenomenon. An admirable example of the intertwining of political and economic power may be found in the British position in Egypt. British economic penetration in Egypt in the last two decades of the nineteenth century resulted from British military occupation, which was designed to protect British interests in the Suez Canal, which had been acquired to protect British trade routes and strategic lines of communication.

The methods used to encourage exports and capture foreign markets are too familiar to need discussion. The simplest of all is the granting of loans or credits to finance exports. Before 1914 Great Britain was so little preoccupied with the problem of markets that loans obtained in London by foreign borrowers were free of any condition as to where the proceeds should be spent. Foreign loans obtained elsewhere frequently carried the condition that the whole or part of the proceeds should be expended by the borrower in the lending country.[1] Since 1919 this condition has been almost universally applied. In Great Britain, two governmental institutions — the Colonial Development Fund and the Export Credits Guarantee Department — were engaged in financing British exports, the first to the Empire, the second to foreign countries. Before 1939, the operations of the Export Credits Guarantee Department were officially described as being of a purely commercial character. But by an act passed in 1939, the limit of the guarantees which might be given by the Department was increased, and a sum of £10,000,000 was earmarked for the guaranteeing of transactions " in connection with which it appears to them [i.e. the Board of Trade] expedient in the national interest that guarantees should be given ".[2] In introducing this measure into the House of Commons, the President of the Board of Trade denied the suggestion that Great Britain " had declared a trade war upon Germany ", but described the measure as one of " economic rearmament " and added that " the economic rearmament which we are trying now to undertake is exactly like our other rearmament ".[3] In July 1939, the amount of £10,000,000 was increased to £60,000,000. Export bounties and currency manipulation are merely indirect forms of export credits.

The most characteristic modern method of acquiring markets and the political power which goes with them is, however, the reciprocal trade agreement — the return to a system of thinly disguised barter. Thus British purchases of meat and cereals in the Argentine and of bacon and butter in Denmark and the Baltic States secured markets in those countries for British coal and British manufactures. The Ottawa Agreements were a slightly more complicated variation on the same theme. In the Central European and Balkan countries Germany, by purchasing local products (mainly cereals and tobacco) for which no other lucrative outlet could be found, secured not only a market for German goods, but a sphere of political influence. One of the symptoms of the artificial character of French political influence in this region

---

[1] Examples from France and Austria are quoted by C. K. Hobson, *The Export of Capital* (1914), p. 16. Russia and Belgium were also lending countries which commonly imposed this condition.

[2] It may be significant that in 1938 an official of the Foreign Office was transferred to the staff of the Export Credits Guarantee Department.

[3] House of Commons, December 15, 1938 : *Official Report*, col. 2319.

was failure to secure any substantial share in its trade. Purchasing power had become an international asset ; and the fact that price was no longer the dominant factor (Germany made most of her purchases in South-Eastern Europe at rates above world prices) put the purchaser and not the producer in a position to call the tune. A new power has thus been placed in the hands of countries with a large population and a high standard of living. But it is a wasting asset which, if used to excess, tends to destroy itself.

Economic Power and International Morality

One concluding reflexion may round off this summary sketch of the use of the economic weapon as an instrument of political power. The substitution of the economic weapon for the military weapon — what Marx calls the replacement of cannons by capital — is a symptom not so much of superior morality as of superior strength. This can be seen from a few simple examples. Great Britain, aggrieved by the trial of the Metro-Vickers engineers in Moscow, could obtain satisfaction by imposing an embargo on Soviet imports. Italy, aggrieved by the murder of an Italian officer, could not avail herself of this economic expedient (for an Italian embargo on Greek imports would have been negligible) ; she could obtain satisfaction only by the brutal military method of bombarding Corfu. In 1931, Great Britain established what came to be known as a " sterling *bloc* " by methods which were non-political and in appearance largely fortuitous. Germany, in order to establish an equivalent " mark *bloc* " in Central and South-Eastern Europe, had to resort to methods which were frankly political and included the use and threatened use of force. British economic and financial strength enabled Great Britain to refrain from intervention in the Spanish civil war. The British Government relied on " sterling bullets " to prevent the permanent predominance of Germany and Italy in Spain, whatever the issue of the war. As regards the Far East, the Prime Minister in the same period remarked that " when the war is over, and the reconstruction of China begins, she cannot possibly be reconstructed without some help from this country ".[1]

[1] House of Commons, November 1, 1938, reprinted in N. Chamberlain, *The Struggle for Peace*, p. 340.

The growing strength of the United States in international trade and finance was one, at any rate, of the reasons which allowed the United States Government to abandon its traditional practice of landing marines in the territory of recalcitrant Latin American republics and to adopt the " good neighbour " policy.

The point, however, has a wider application to the whole problem of " aggression " and territorial annexation. One of the most revealing documents on this aspect of power is a despatch from the Russian Chargé d'Affaires in Peking to the Russian Government in 1910 :

> Should we be sufficiently powerful economically [wrote this frank diplomat], it would be simpler to direct all our efforts to the conclusion of an economic treaty. If, however, as I fear, we should by so doing only be of service to foreigners and ourselves be unable to secure any profits from what had been achieved (thus we have for instance in reality been unable to profit by the extraordinary advantages embodied in the commercial treaty of 1881), then there is, in my opinion, no reason to depart from the basis of policy we have followed hitherto, that of territorial acquisition.[1]

A recent British writer on the Far East has made a similar observation :

> Free Trade, as championed by England in the nineteenth century, was the cause of the stronger in purely commercial competition. The " sphere of influence " with its special rights was the objective of states which sought to compensate for weakness in such competition by the direct application of political power.[2]

Great Britain's unchallenged naval and economic supremacy throughout the nineteenth century enabled her to establish a commanding position in China with a minimum of military force and of economic discrimination. A relatively weak Power like Russia could only hope to achieve a comparable result by naked aggression and annexation. Japan afterwards learned the same lesson. In his well-known memorandum of January 1907, Crowe argued that Great Britain was " the natural protector of the weaker communities ", and that by her free-trade policy of an open market " she undoubtedly strengthens her

---

[1] B. de Siebert, *Entente Diplomacy of the World War*, p. 20.
[2] G. F. Hudson, *The Far East in World Politics*, p. 54.

hold on the interested friendship of other nations ".[1] The
argument might have been developed by adding that Great
Britain, in virtue of her inherent economic strength and the
free-trade policy made possible by it, was able to exercise in
many countries a measure of indirect influence and control
which no other Power could have achieved without interference
with the political independence of the countries concerned, and
that this advantage made it as natural for Great Britain, as
it would have been difficult for others, to appear as a champion
of the political independence of small nations. In Egypt,
Great Britain has reconciled her military and economic predominance with the formal independence of the country, where
a weaker Power would have had to resort to annexation to
obtain a similar effect. Great Britain was able to abandon
her formal authority over Iraq and to maintain her interests
there, while France shrank from the same step in Syria. The
economic weapon is pre-eminently the weapon of strong Powers.
It is significant that a proposal made by the Soviet Government
in 1931 for a pact of " economic non-aggression " was received
with the greatest hostility by the three most powerful countries
of the day : Great Britain, France and the United States.

Nevertheless, it is perhaps difficult to dismiss as unfounded
the common view that the use of the economic weapon is less
immoral than the use of the military weapon. This may not
always be true. Blockade in time of war may cause as much
suffering as a series of air raids. But generally speaking, there
is a sense in which dollars are humaner than bullets even if
the end pursued be the same. It is less immoral to place an
embargo on Soviet imports than to bombard Greeks. It cannot
be reasonably doubted that a form of economic control (such
as that of the United States in Central America) which preserves
a measure of political independence is more acceptable to
subordinate nations, and therefore less immoral, than direct
political control (such as that established by Germany in 1939
in Bohemia and Moravia). The distinction is not entirely
removed by pointing out that the United States, if she were
economically as weak as Germany, might well have taken the
same course. It is true that the poor are more likely to steal
than the rich, and that this affects our moral judgment of

[1] *British Documents on the Origins of the War,* ed Gooch and Temperley,
iii. p. 403.

individual cases of theft. But theft is generally recognised as *per se* immoral. This is merely an illustration of the way in which morality itself is involved in questions of power.

The moral issue will require consideration later. For the present, the most important lesson to be drawn in this field is the illusory character of the popular distinction between economic and military power. Power, which is an element of all political action, is one and indivisible. It uses military and economic weapons for the same ends. The strong will tend to prefer the minor and more " civilised " weapon, because it will generally suffice to achieve his purposes ; and as long as it will suffice, he is under no temptation to resort to the more hazardous military weapon. But economic power cannot be isolated from military power, nor military from economic. They are both integral parts of political power ; and in the long run one is helpless without the other.

(c) POWER OVER OPINION

Power over opinion is the third form of power. The " Jingoes " who sang " We've got the ships, we've got the men, we've got the money too " had accurately diagnosed the three essential elements of political power : armaments, manpower and economic power. But man-power is not reckoned by mere counting of heads. " The Soldan of Egypt or the Emperor of Rome ", as Hume remarked, " might drive his harmless subjects like brute beasts against their sentiments and inclinations. But he must at least have led his *mamelukes* or pretorian bands like men by their opinions." [1] Power over opinion is therefore not less essential for political purposes than military and economic power, and has always been closely associated with them. The art of persuasion has always been a necessary part of the equipment of a political leader. Rhetoric has a long and honoured record in the annals of statesmanship. But the popular view which regards propaganda as a distinctively modern weapon is, none the less, substantially correct.

Propaganda in the Modern World

The most obvious reason for the increasing prominence attached to power over opinion in recent times is the broadening

---

[1] *The Philosophical Works of David Hume*, iv p. 31.

of the basis of politics, which has vastly increased the number of those whose opinion is politically important. Until comparatively modern times, those whose opinion it was worth while to influence were few in number, united by close ties of interest and, generally speaking, highly educated ; and the means of persuasion were correspondingly limited. " Scientific exposition ", in Hitler's words, is for the intelligentsia. The modern weapon of propaganda is for the masses.[1] Christianity seems to have been the first great movement in history with a mass appeal. Appropriately enough, it was the Catholic church which first understood and developed the potentialities of power over large masses of opinion. The Catholic church in the Middle Ages was — and has, within the limits of its power, remained — an institution for diffusing certain opinions and extirpating other opinions contrary to them : it created the first censorship and the first propaganda organisation. There is much point in the remark of a recent historian that the mediaeval church was the first totalitarian state.[2] The Reformation was a movement which simultaneously deprived it, in several parts of Europe, of its power over opinion, of its wealth and of the authority which the military power of the Empire had conferred on it.

The problem of power over opinion in its modern mass form has been created by developments in economic and military technique — by the substitution of mass-production industries for individual craftsmanship and of the conscript citizen army for the volunteer professional force. Contemporary politics are vitally dependent on the opinion of large masses of more or less politically conscious people, of whom the most vocal, the most influential and the most accessible to propaganda are those who live in and around great cities. The problem is one which no modern government ignores. In appearance, the attitude adopted towards it by democracies and by totalitarian states is diametrically opposed. Democracies purport to follow mass opinion ; totalitarian states set a standard and enforce conformity to it. In practice, the contrast is less clear cut. Totalitarian states, in determining their policy, profess to express the will of the masses ; and the profession is not wholly vain. Democracies, or the groups which control them, are not

[1] Hitler, *Mein Kampf*, p. 196.
[2] G. G. Coulton, *Mediaeval Panorama*, p. 458 *et al.*

altogether innocent of the arts of moulding and directing mass opinion. Totalitarian propagandists, whether Marxist or Fascist, continually insist on the illusory character of the freedom of opinion enjoyed in democratic countries. There remains a solid substratum of difference between the attitude of democracies and totalitarian states towards mass opinion, which may prove a decisive factor in times of crisis. But both agree in recognising its paramount importance.

The same economic and social conditions which have made mass opinion supremely important in politics have also created instruments of unparalleled range and efficiency for moulding and directing it. The oldest, and still perhaps the most powerful, of these instruments is universal popular education. The state which provides the education necessarily determines its content. No state will allow its future citizens to imbibe in its schools teaching subversive of the principles on which it is based. In democracies, the child is taught to prize the liberties of democracy ; in totalitarian states, to admire the strength and discipline of totalitarianism. In both, he is taught to respect the traditions and creeds and institutions of his own country, and to think it better than any other. The influence of this early unconscious moulding is difficult to exaggerate. Marx's dictum that " the worker has no country " has ceased to be true since the worker has passed through national schools.

But when we speak of propaganda to-day, we think mainly of those other instruments whose use popular education has made possible : the radio, the film and the popular press. The radio, the film and the press share to the fullest extent the characteristic attribute of modern industry, i.e. that mass-production, quasi-monopoly and standardisation are a condition of economical and efficient working. Their management has, in the natural course of development, become concentrated in fewer and fewer hands ; and this concentration facilitates and makes inevitable the centralised control of opinion. The mass-production of opinion is the corollary of the mass-production of goods. Just as the nineteenth-century conception of political freedom was rendered illusory for large masses of the population by the growth and concentration of economic power, so the nineteenth-century conception of freedom of thought is being fundamentally modified by the development of these new and extremely powerful instruments of power over opinion. The prejudice which the word propaganda still excites in many minds to-day [1] is closely parallel to the prejudice against state control of industry and trade. Opinion, like trade and industry, should according to the old liberal conception be allowed to flow in its own natural channels without artificial regulation. This conception has broken down on the hard fact that in modern conditions opinion, like trade, is not and cannot be exempt from artificial controls. The issue is no longer whether men shall be politically free to express their opinions, but whether freedom of opinion has, for large masses of people, any meaning but subjection to the influence of innumerable forms of propaganda directed by vested interests of one kind or another. In the totalitarian countries, radio, press and film are state industries absolutely controlled by governments. In democratic countries, conditions vary, but are everywhere tending in the direction of centralised control. Immense corporations are called into existence, which are too powerful and too vital to the community to remain wholly independent of the machine of government, and which themselves find it convenient to accept voluntary collaboration with the state as an alternative to formal control by it. The nationalisation of opinion has proceeded everywhere *pari passu* with the nationalisation of industry.

Propaganda as an Instrument of Policy

The organised use of power over opinion as a regular instrument of foreign policy is a modern development. Before 1914, cases occurred of the use of propaganda by governments in international relations. The press was freely used by Bismarck and other statesmen, though rather for the purpose of making pronouncements to foreign governments than as a means of influencing public opinion at large. Co-operation between the missionary and the trader, and the support of both by military force, was a familiar nineteenth-century example of unofficial association between propaganda and economic and military power in the interests of national expan-

[1] " I wish ", said the Home Secretary in the House of Commons on July 28, 1939, " there had been no necessity for any Government publicity anywhere in the world. I still look forward to living long enough to see an end of this objectionable relic of the years of the war " (*Official Report*, col. 1834).

sion. But the field of propaganda was limited ; and the only people who exploited it at all intensively were the revolutionaries. Any systematic resort to propaganda by governments would have been thought undignified and rather disreputable.

It did not take long for the belligerents of 1914–18 to realise that " psychological war must accompany economic war and military war ".[1] It was a condition of success on the military and economic fronts that the " morale " of one's own side should be maintained, and that of the other side sapped and destroyed. Propaganda was the instrument by which both these ends were pursued. Leaflets were dropped over the enemy lines inciting his troops to mutiny ; and this procedure, like most new weapons of war, was at first denounced as being contrary to international law.[2] Moreover, the new conditions of warfare nullified, in this as in so many other respects, the distinction between combatant and civilian ; and the morale of the civilian population became for the first time a military objective.

> Long-distance bombing [wrote the British Chief of Staff in January 1918] will produce its maximum moral effect only if visits are constantly repeated at short intervals so as to produce in each area bombed a sustained anxiety. It is this recurrent, as opposed to isolated spasmodic attacks, which interrupts industrial production and undermines public confidence.[3]

The military chiefs of other belligerent countries were doubtless considering the same problem in similar terms. The demoralisation of the civilian population was the primary objective not only of many air raids but of the German long-range bombardment of Paris by " big Bertha " ; and the work of the bomb and the shell was reinforced, especially during the last months of the war, by an intense output of printed propaganda. Throughout the first world war the close interdependence

[1] H. D. Lasswell in the Foreword to G. G. Bruntz, *Allied Propaganda and the Collapse of the German Empire.* This book is the most comprehensive available account of its subject.

[2] In 1917, two British airmen captured by the Germans were sentenced to ten years' hard labour for dropping such leaflets in contravention of the laws of war. The sentences were remitted on a British threat of reprisals. The practice was explicitly sanctioned in The Hague rules of 1923 for the conduct of aerial warfare (Bruntz, *op. cit.* pp. 142-4).

[3] *The War in the Air* (British Official History of the War), by H. A. Jones vi. Appendix VI, p. 26.

between the three forms of power was constantly demonstrated. The success of propaganda on both sides, both at home and in neutral and enemy countries, rose and fell with the varying fortunes of the military and economic struggle. When at length the Allied blockade and Allied victories in the field crippled German resources, Allied propaganda became enormously effective and played a considerable part in the final collapse. The victory of 1918 was achieved by a skilful combination of military power, economic power, and power over opinion.

Notwithstanding the general recognition of the importance of propaganda in the later stages of the war, it was still regarded by almost everyone as a weapon specifically appropriate to a period of hostilities. " In the same way as I send shells into the enemy trenches, or as I discharge poison gas at him," wrote the German general who was primarily responsible for despatching Lenin and his party in the sealed train to Russia, " I, as an enemy, have the right to use propaganda against him." [1] The abolition of ministries and departments of propaganda at the end of the war was an automatic measure of demobilisation. Yet within twenty years of the armistice, in what was still formally a time of peace, many governments were conducting propaganda with an intensity unsurpassed in the war period ; and new official or semi-official agencies for the influencing of opinion at home and abroad were springing up in every country. This new development was rendered possible and inevitable by the popularisation of international politics and by the growing efficiency of propaganda methods. Since both these processes are likely to continue, its permanence seems assured.

The initiative in introducing propaganda as a regular instrument of international relations must be credited to the Soviet Government. The causes of this were partly accidental. The Bolsheviks, when they seized power in Russia, found themselves desperately weak in the ordinary military and economic weapons of international conflict. The principal element of strength in their position was their influence over opinion in other countries ; and it was therefore natural and necessary that they should exploit this weapon to the utmost. In early days, they seriously believed in their ability to dissolve

---

[1] Hoffmann, *War Diaries* (Engl. transl.), ii. p. 176.

the German armies by the distribution of propaganda leaflets and by fraternisation between the lines. Later, they counted on propaganda in Allied countries to paralyse Allied intervention against them in the civil war. Had not propaganda been supplemented by the creation of an effective Red Army, it might by itself have proved ineffective. But the importance of the role it played is sufficiently indicated by the fear of Bolshevik propaganda felt for many years afterwards, and not yet extinct in many European and Asiatic countries. Soviet Russia was the first modern state to establish, in the form of the Communist International, a large-scale permanent international propaganda organisation.

There was, however, a profounder cause why control over opinion should have taken a foremost place in the policy of Soviet Russia. Since the end of the Middle Ages, no political organisation had claimed to be the repository of universal truth or the missionary of a universal gospel. Soviet Russia was the first national unit to preach an international doctrine and to maintain an effective world-propaganda organisation. So revolutionary did this innovation appear that the Communist International purported at the outset to be wholly unconnected with the power of the Soviet Government. But this separation, which may have been effective in details of administration, never extended to major issues of policy ; and after the Soviet state had been consolidated under Stalin, the separation became no more than a polite fiction. This development had far more than a local significance, and gives us the clue to the whole problem of the place of what are now known as " ideologies " in international politics. For if it be true that power over opinion cannot be dissociated from other forms of power, then it appears to follow that, if power cannot be internationalised, there can be no such thing in politics as international opinion, and international propaganda is as much a contradiction in terms as an international army. This view, paradoxical as it may appear, can be supported by extremely cogent arguments ; and both it and its implications require careful examination.

National or International Propaganda ?

Most political ideas which have strongly influenced mankind have been based on professedly universal principles and have therefore had, at any rate in theory, an international character. The ideas of the French Revolution, free trade, communism in its original form of 1848 or in its reincarnation of 1917, Zionism, the idea of the League of Nations, are all at first sight (as they were in intention) examples of international opinion divorced from power and fostered by international propaganda. But reflexion will set limits on this first impression. How far were any of these ideas politically effective until they took on a national colour and were supported by national power ? The answer is not easy. Albert Sorel has a well-known passage on the course taken by the enthusiasm of the French revolutionaries :

> They confuse . . . the propagation of the new doctrines with the extension of French power, the emancipation of mankind with the greatness of the Republic, the rule of reason with that of France, the liberation of peoples with the conquest of states, the European revolution with the domination of the French Revolution over Europe.[1]

The military power of Napoleon was notoriously the most potent factor in the propagation throughout Europe of the ideas of 1789. The political influence of the idea of free trade dated from its adoption by Great Britain as the basis of British policy. The revolutionaries of 1848 failed everywhere to achieve political power ; and the ideas of 1848 remained barren. Neither the First nor the Second International attained any real authority. As 1914 shewed, there were national labour movements, but there was no international labour movement. The Third or Communist International enjoyed little influence until the power of the Russian state was placed behind it ; and Stalin has garbled and disseminated the ideas of 1917 in much the same way as Napoleon garbled and disseminated the ideas of 1789. Trotskyism, unsupported by the power of any state, remains without influence. Zionism, politically impotent so long as it relied solely in international propaganda, is effective in so far as it can count on the political backing of Great Powers. Propaganda is ineffective as a political force until it acquires a national home and becomes linked with military and economic power.

The fate of the League of Nations and of propaganda on its

[1] A. Sorel, *L'Europe et la Révolution Française*, pp. 541-2.

behalf is perhaps the best modern illustration of this tendency. As has been shewn, men like Woodrow Wilson and Lord Cecil conceived the League of Nations as an expression of " the organised opinion of mankind " controlling the military and economic power of governments. International public opinion was the supreme instrument of power (" by far the strongest weapon we have ") ; and this opinion was to be created by international propaganda which took no heed of frontiers.[1] Throughout the nineteen-twenties, this fallacy of the power of international opinion was being gradually exposed. That it survived at all was due to the persistent use by League enthusiasts of slogans like peace and disarmament which were capable of a universal appeal precisely because they meant different, and indeed contradictory, things to different people. Every country wanted to achieve the aims of its policy without war, and therefore stood for peace. Every country wanted disarmament of other countries or disarmament in those weapons which it did not regard as vital to itself. After the collapse of the Disarmament Conference, it became apparent to all that the League of Nations could be effective only in so far as it was an instrument of the national policy of its most powerful members. Opinion in favour of the League ceased altogether to be international, and was confined to those countries where the League was felt to be serving ends of national policy. In Great Britain the League of Nations became for the first time popular with what might be called the nationalist wing of the Conservative Party.

The fallacy of belief in the efficacy of an international public opinion divorced from national power may be further illustrated by developments elsewhere. The group of movements conveniently classified under the rubric of Fascism was based on certain professedly universal principles such as the rejection of democracy and class-warfare, the insistence on leadership, and so forth. In its early days, Fascism was authoritatively described as " not an article for export ", and was for many years so treated by the countries which adhered to it. At a later date this limitation was explicitly disclaimed,[2] and Fascism became the theme of a vigorous international propaganda in many parts of the world. It would, however, be a superficial

---

[1] See pp. 31-36.
[2] Mussolini, *Scritti e Discorsi*, vi. 151 ; vii. 230.

diagnosis to pretend that, while the League of Nations and the Communist International began as instruments of international opinion and ended as instruments of national policy, Fascism began as an instrument of national policy and ended as an instrument of international opinion. In both cases, the international phase was an illusion (which does not mean that many people may not sincerely believe in it). International propaganda for Fascism was an instrument of the national policy of certain states, and grew with the growth of the military and economic power of those states. But the *reductio ad absurdum* of international ideological propaganda as a cloak for national policy came with the adoption of negative slogans designed to unite in a political alliance those who shared no positive ideology in common. Thus the Anti-Comintern Pact did not prevent Germany from coming to an agreement with the principal Communist Power when the needs of national policy seemed to require it ; and the " anti-Fascism " of the democratic nations did not deter them from seeking the alliance of countries whose forms of government were indistinguishable from Fascism. These slogans had no meaning or substance apart from the national policies of the countries by which they were used. Power over opinion cannot be dissociated from military and economic power.

International Agreements regarding Propaganda

Propaganda is now so well recognised as a national political weapon that stipulations regarding its use are fairly common in international agreements. Such stipulations were, appropriately enough, first introduced into agreements made with the Soviet Government for the purpose of limiting the activities of the Communist International. But this could still be thought of as an exceptional case. Outside Soviet Russia, the first recorded agreement to abstain from hostile propaganda seems to have been one concluded between the German and Polish Broadcasting Companies, which undertook to assure that " the matter broadcast does not in any way offend the national sentiment of listeners who are nationals of the other contracting party ".[1] Propaganda was first raised to the dignity of a universal issue when the Polish Government made proposals

[1] *League of Nations*, C.602, M.240, 1931, ix. p. 4.

to the Disarmament Conference for a convention on "moral disarmament" To limit the propaganda weapon by a general convention proved as hopeless a task as to limit the military weapon.[1] But bilateral agreements for terminating hostile propaganda were concluded between Germany and Poland in 1934 and between Germany and Austria in 1936;[2] and in the Anglo-Italian Agreement of April 16, 1938, the two countries "placed on record their agreement that any attempt by either of them to employ the methods of publicity and propaganda at its disposal in order to injure the interests of the other would be inconsistent with the good relations which it is the object of the present agreement to establish ".

Such agreements create an obvious difficulty for democracies, which purport not to limit the free expression and publication of opinions about international affairs, and cannot therefore formally undertake to prevent propaganda on their territory against any country ; and this embarrassment is reflected in the contorted phraseology of the Anglo-Italian Agreement. The fact is, however, that in the sphere of opinion, as in the economic sphere, the nineteenth-century principles of *laissez-faire* no longer hold good, even for democracies. Just as democratic governments have been compelled to control and organise economic life in their territories in order to compete with totalitarian states, so they find themselves at a disadvantage in dealing with these states if they are not in a position to control and organise opinion. Recognition of this fact grew rapidly even in Great Britain. In questions affecting international relations, a discreet influence, amounting in times of crisis to direct though unofficial censorship, was exercised even before the outbreak of the second world war over broadcasting, films and press ; and though the use of this influence was frequently criticised in particular cases, it became clear that some such

[1] An international convention under which the parties undertook to prevent the broadcasting from their territories of "incitements to war", or in general hostile propaganda, against other contracting parties, was signed at Geneva by most of the surviving members of the League in September 1936 (*League of Nations*, C.399 (1), M.252 (1), 1936, xii.).

[2] In both cases the agreement about propaganda did not figure in an officially published text, but its existence was disclosed in communiqués. The communiqué of the Austrian Foreign Office on the German-Austrian Agreement of July 11, 1936, announced that " both countries are to refrain from all aggressive uses of the wireless, films, news services and the theatre " (*Documents on International Affairs, 1936*, p. 324).

measures of restraint would be applied in similar circumstances by whatever government happened to be in power.[1] Simultaneously, there was a rapid extension of propaganda designed to familiarise foreign opinion with the British point of view. Since 1935, a body called the British Council has exercised the function of "making the life and thought of the British peoples more widely known abroad" In 1938, the British Broadcasting Corporation began the regular broadcasting of news bulletins in various foreign languages. In June 1939, the Prime Minister announced the creation of a new Foreign Publicity Department of the Foreign Office, which served as a nucleus for the Ministry of Information set up immediately on the outbreak of war.

Truth and Morality in Propaganda

We have hitherto discussed power over opinion in precisely the same terms as military and economic power ; and the close connexion between these different forms of power is so vital, and has been so much neglected in theoretical discussion, that this seems the most fruitful approach to the problem at the present time. Some people might indeed argue that this is the only correct approach. For in the first place, opinion is conditioned by status and interest ; and secondly, as we have seen in a previous chapter, a ruling class or nation, or dominant group of nations, not only evolves opinions favourable to the maintenance of its privileged position, but can, in virtue of its military and economic superiority, easily impose these opinions on others. The victory of the democratic countries in 1918 created an almost universal opinion that democracy was the best form of government. In the nineteen-thirties, opinion in many parts of the world on the merits of Fascism as a form of government may be said, without much exaggeration, to have varied *pari passu* with the military and economic power of Germany and Italy in relation to the other Great Powers.

[1] A revealing debate on the press, initiated by the Liberal Opposition, took place in the House of Commons on December 7, 1938. While Liberal speakers argued for the freedom of the press on familiar nineteenth-century lines, the spokesman of the Labour Opposition declared that the freedom of the press was already illusory, and wanted to " make every newspaper in the country responsible for every item of news it prints and answerable to this House or some public authority " (*Official Report*, col. 1293).

These propositions could be supported by innumerable examples. If they were absolutely true, then power over opinion would in fact be indistinguishable in character from military and economic power, and there would be nothing which, given sufficient power and technical skill, men could not be made to believe. That this is the case has indeed sometimes been suggested. " By clever, persistent propaganda ", said Hitler, " even heaven can be represented to a people as hell, and the most wretched life as paradise " ; [1] and American advertising specialists are alleged to hold that " only cost limits the delivery of public opinion in any direction on any topic ".[2] But these are the pardonable exaggerations of expert practitioners. As we shall see, even Hitler did not really believe in the unlimited power of propaganda to manufacture opinion. Here as elsewhere, the extreme realist position becomes untenable. When we set power over opinion side by side with military and economic power, we have none the less to remember that we are dealing no longer with purely material factors, but with the thoughts and feelings of human beings.

Absolute power over opinion is limited in two ways. In the first place, it is limited by the necessity of some measure of conformity with fact. There are objective facts which are not totally irrelevant to the formation of opinion. Good advertising may persuade the public that a face cream made of inferior materials is the best. But the most expert advertiser could not sell a face cream made of vitriol. Hitler condemned the futility of German propaganda in the first world war which depicted the enemy as ridiculous and contemptible. The propaganda was unsuccessful simply because it was, as the German soldier in the trenches discovered, untrue. This danger that " truth will out ", especially in an age of competitive propaganda, is a serious limitation on power over opinion. Education, which is one of the strongest instruments of this power, tends at the same time to promote a spirit of independent enquiry which is also one of the strongest antidotes against it. In so far as it strains and interprets facts for a specific purpose, propaganda always contains within itself this potentially self-defeating element.

Secondly, power over opinion is limited — and perhaps even

[1] Hitler, *Mein Kampf*, p. 302.
[2] J. Truslow Adams, *The Epic of America*, p. 360.

more effectively — by the inherent utopianism of human nature. Propaganda, harnessed to military and economic power, always tends to reach a point where it defeats its own end by inciting the mind to revolt against that power. It is a basic fact about human nature that human beings do in the long run reject the doctrine that might makes right. Oppression sometimes has the effect of strengthening the will, and sharpening the intelligence, of its victims, so that it is not universally or absolutely true that a privileged group can control opinion at the expense of the unprivileged. As Hitler himself wrote, " every persecution which lacks a spiritual basis " has to reckon with a " feeling of opposition to the attempt to crush an idea by brute force ".[1] And this vital fact gives us another clue to the truth that politics cannot be defined solely in terms of power Power over opinion, which is a necessary part of all power, can never be absolute. International politics are always power politics ; for it is impossible to eliminate power from them. But that is only part of the story. The fact that national propaganda everywhere so eagerly cloaks itself in ideologies of a professedly international character proves the existence of an international stock of common ideas, however limited and however weakly held, to which appeal can be made, and of a belief that these common ideas stand somehow in the scale of values above national interests. This stock of common ideas is what we mean by international morality.

[1] Hitler, *Mein Kampf*, p. 187.

CHAPTER 9

MORALITY IN INTERNATIONAL POLITICS

THE place of morality in international politics is the most obscure and difficult problem in the whole range of international studies. Two reasons for its obscurity, one general and one particular, may be suggested.

In the first place, most discussions about morality are obscured by the fact that the term is commonly used to connote at least three different things :

(i) The moral code of the philosopher, which is the kind of morality most rarely practised but most frequently discussed.

(ii) The moral code of the ordinary man, which is sometimes practised but rarely discussed (for the ordinary man seldom examines the moral assumptions which underlie his actions and his judgments and, if he does, is peculiarly liable to self-deception).

(iii) The moral behaviour of the ordinary man, which will stand in fairly close relation to (ii), but in hardly any relation at all to (i).

It may be observed that relationship between (ii) and (iii) is mutual. Not only is the behaviour of the ordinary man influenced by his moral code, but his moral code is influenced by the way in which ordinary men, including himself, behave. This is particularly true of the ordinary man's view of political morality, which tends, more than personal morality, to be a codification of existing practice, and in which the expectation of reciprocity always plays an important part.

The monopoly of international studies between the two wars by the utopian school resulted in a concentration of interest on discussions of the question what international morality ought ideally to be. There was little discussion of the moral behaviour of states except to pass hasty and sweeping condemnation on it in the light of this ideal morality. There was no discussion at all of the assumptions of the ordinary man about international morality. This was particularly unfortunate at a period in which the popularisation of politics for the first time made the assumptions of the ordinary man a matter of primary importance ; and the ever widening rift between the international utopia and international reality might have been described in terms of this divergence between the theory of the philosopher and practice based on the unexpressed and often unconscious assumptions of the ordinary man. Moreover, utopia met its usual fate in becoming, unknown to itself, the tool of vested interests. International morality, as expounded by most contemporary Anglo-Saxon writers, became little more than a convenient weapon for belabouring those who assailed the *status quo*. Here as elsewhere, the student of international politics cannot wholly divest himself of utopianism. But he will be well advised to keep his feet on the ground and rigorously maintain contact between his ambitions for the future and the realities of the present. Nor should this be too difficult. The anthropologist who investigates the moral codes and behaviour of a cannibal tribe probably starts from the presupposition that cannibalism is undesirable, and is conscious of the desire that it should be abolished. But he may well be sceptical of the value of denunciations of cannibalism, and will in any case not mistake such denunciations for a scientific study of the subject. The same clarity of thought has not always distinguished students of international morality, who have generally preferred the role of the missionary to that of the scientist

The second obscurity is peculiar to the international field. Strange as it may appear, writers on international morality are not agreed among themselves — and are not always clear in their own minds — whether the morality which they wish to discuss is the morality of states or the morality of individuals. This point is so vital to the whole discussion that it must be cleared up on the threshold of our enquiry.

The Nature of International Morality

The period of absolute personal rule in which the modern state first began to take shape was not much troubled by distinction between personal and state morality. The personal responsibility of the prince for acts of state could be assumed without any undue straining of the facts. Charles I may have been a good father and a bad king. But in both capacities, his acts could be treated as those of an individual.[1] When, however, the growing complication of the state machine and the development of constitutional government made the personal responsibility of the monarch a transparent travesty, the personality (which seemed a necessary condition of moral responsibility) was transferred from the monarch to the state. Leviathan, as Hobbes said, is an " Artificial Man ". This was an important step forward. It was the personification of the state which made possible the creation of international law on the basis of natural law. States could be assumed to have duties to one another only in virtue of the fiction which treated them as if they were persons. But the personification of the state was a convenient way of conferring on it not merely duties, but rights ; and with the growth of state power in the nineteenth and twentieth centuries state rights became more conspicuous than state duties. Thus the personification of the state, which began as a liberal and progressive device, came to be associated with the assertion of unlimited rights of the state over the individual and is now commonly denounced as reactionary and authoritarian. Modern utopian thinkers reject it with fervour,[2] and are consequently led to deny that morality can be attributed to the state. International morality must, on this view, be the morality of individuals.

The controversy about the attribution of personality to the state is not only misleading, but meaningless. To deny personality to the state is just as absurd as to assert it. The personality of the state is not a fact whose truth or falsehood is a matter for argument. It is what international lawyers have called " the postulated nature " of the state.[3] It is a necessary fiction or hypothesis — an indispensable tool devised by the human

[1] The Allied Governments in the Versailles Treaty attempted to revive this historic assumption by holding the ex-Kaiser personally responsible for acts of state ; but the attempt was almost universally condemned as soon as passions began to cool. Modern dictatorships, however, helped to bring this conception back to fashion. Thus Professor Toynbee called the invasion of Abyssinia " Signor Mussolini's deliberate personal sin " (*Survey of International Affairs*, *1935*, ii. p. 3), though he would probably have felt it incongruous to describe the Hoare-Laval Plan as the " personal sin " of Sir S. Hoare or Laval.

[2] Duguit, for example, calls it "valueless and meaningless anthropomorphism" (*Traité de droit constitutionnel*, i. ch. v.).

[3] Hall, *International Law* (8th ed.), p. 50 ; Pearce Higgins, *International Law and Relations*, p. 38.

mind for dealing with the structure of a developed society.[1] It is theoretically possible to imagine a primitive political order in which individuals are individuals and nothing more, just as it is possible to imagine an economic order in which all producers and traders are individuals. But just as economic development necessitated resort to the fiction of corporate responsibility in such forms as that of the joint-stock company, so political development necessitated the fiction of the corporate responsibility of the state. Nor are the rights and obligations of these fictitious entities regarded as purely legal. A bank is praised for generosity to its employees, an armaments firm is attacked for unpatriotic conduct, and railways have "obligations to the public" and demand a "square deal" — all issues implying the relevance, not merely of legal, but of moral standards. The fiction of the group-person, having moral rights and obligations and consequently capable of moral behaviour, is an indispensable instrument of modern society ; and the most indispensable of these fictitious group-persons is the state. In particular, it does not seem possible to discuss international politics in other terms. " Relations between Englishmen and Italians " is not a synonym for " relations between Great Britain and Italy ". It is a curious and significant paradox that those utopian writers on international affairs who most vigorously denounce the personification of the state as absurd and sinister none the less persistently allocate moral praise and blame (generally the latter) to those imaginary entities, " Great Britain ", " France " and " Italy ", whose existence they deny.

Continuity is another element in society which makes the fiction of the group-person indispensable. The keenest objectors to the personification of the state will have no qualms about celebrating the 150th anniversary of *The Times* or the 38th victory of " Cambridge " in the boat race, and will confidently expect " the London County Council " to repay, fifty years hence, money which " it " borrows and spends to-day. Personification is the category of thought which expresses the continuity of institutions ; and of all institutions the state is

---

[1] This does not, of course, mean that the state is a necessary form of political organisation, but only that, so long as the state *is* the accepted form, its personification is a necessary fiction. The same would apply to any other form (e.g. the class). The personification of the proletariat has gone far in Soviet Russia (e.g. the fiction that it " owns " the means of production).

the one whose continuity it is most essential to express. The question whether the Belgian Guarantee Treaty of 1839 imposed an obligation on Great Britain to assist Belgium in 1914 raised both legal and moral issues. But it cannot be intelligently discussed except by assuming that the obligation rested neither personally on Palmerston who signed the treaty of 1839, nor personally on Asquith and Grey who had to decide the issue in 1914, neither on all individual Englishmen alive in 1839, nor on all individual Englishmen alive in 1914, but on that fictitious group-person " Great Britain ", which was regarded as capable of moral or immoral behaviour in honouring or dishonouring an obligation.[1] In short, international morality is the morality of states. The hypothesis of state personality and state responsibility is neither true nor false, because it does not purport to be a fact, but a category of thought necessary to clear thinking about international relations. It is true that another moral issue was also raised in 1914 — the obligation of individual Englishmen. But this was an obligation to " Great Britain ", arising out of the obligation of " Great Britain " to " Belgium ". The two obligations were distinct ; and confused thinking is the inevitable penalty of failure to distinguish between them.

Curiously enough, this distinction seems to present more difficulty to the philosopher than to the ordinary man, who readily distinguishes between the obligation of the individual to the state, and the obligation of the state to another state. In 1935, the Opposition in the House of Commons denounced the Hoare-Laval Plan as " a terrible crime ". But it did not denounce Sir S. Hoare as a criminal or regard him as such ; it found him guilty only of an error of judgment. In 1938, some Englishmen felt " ashamed " of the Munich Agreement. They were not " ashamed " of themselves ; for they would have done anything in their power to prevent it. They were

[1] A striking example of confused thinking on this subject occurred in a recent letter to *The Times*. Commenting on the alleged British obligation to France in 1914, a distinguished professor of history wrote that " Grey may have regarded his personal honour as involved in support of France, but he certainly did not think that of the Cabinet was " (*The Times*, February 28, 1939). The promise, if any, to support France must have been given by Grey not on his own behalf, but on behalf of Great Britain. Unless he believed that the whole Cabinet was under the same obligation as himself to see that Great Britain's promise was honoured, he could not properly have given it at all.

not " ashamed " of Mr. Chamberlain ; for most of them
admitted that he had acted honestly, though mistakenly, and
one does not feel " ashamed " of anyone who commits an honest
mistake. They were " ashamed " of " Great Britain ", whose
reputation had, in their view, been lowered by a cowardly and
unworthy act. In both these cases, the same act which (in
the view of the critics) represented an intellectual failure on
the part of the individual represented a moral failure on the
part of " Great Britain ". The *mot* became current that the
British loan of £ 10,000,000 to Czecho-Slovakia was "conscience
money ". The essence of " conscience money " is that it is
paid by a moral delinquent ; and the moral delinquent who
paid the £ 10,000,000 was not Mr. Chamberlain, and not
those individual Englishmen who had applauded the Munich
agreement, but " Great Britain ". The obligation of the state
cannot be identified with the obligation of any individual or
individuals ; and it is the obligations of states which are the
subject of international morality.

Two objections are commonly raised to this view.

The first is that the personification of the state encourages the
exaltation of the state at the expense of the individual. This
objection, though it accounts for the disfavour into which the
personification of the state has fallen among liberal thinkers,
is trivial. The personification of the state is a tool ; and to
decry it on the ground of the use to which it is sometimes put
is no more intelligent than to abuse a tool for killing a man.
The tool can equally well be put to liberal uses through emphasis
on the duty of the state both to the individual and to other
states. Nor can democracy altogether dispense with personification as a means to emphasise the duty of the individual.
The most sophisticated of us would probably shrink from paying
taxes to a group of individual fellow-citizens, though we pay
them with comparative alacrity to a personified state. The
same applies with greater force to graver sacrifices. " You
would never have got young men to sacrifice themselves for
so unlucky a country as Ireland ", said Parnell, " only that they
pictured her as a woman." [1] " Who dies if England live ? "
is not adequately paraphrased by " Who dies if other Englishmen live ? " Moreover, it is difficult to see how orderly international relations can be conducted at all unless Englishmen,

[1] Quoted in *Democracy and War*, ed. G. E. C. Catlin, p. 128.

Frenchmen and Germans believe (however absurd the belief may be) that " Great Britain ", " France " and " Germany " have moral duties to one another and a reputation to be enhanced by performing those duties. The spirit of international relations seems more likely to be improved by stimulating this belief than by decrying it. In any case, it is clear that human society will have to undergo a material change before it discovers some other equally convenient fiction to replace the personification of the political unit.

The second objection is more serious. If international morality is the morality of fictitious entities, is it not itself fictitious and unreal ? We can at once accept the view that moral behaviour can only proceed from individuals. To deny that " relations between Great Britain and Italy " means the same as " relations between Englishmen and Italians " is not to deny that " relations between Great Britain and Italy " depend on the actions of individual Englishmen and Italians. The moral behaviour of the state is a hypothesis ; but we need not regard as " unreal " a hypothesis which is accepted in certain contexts as a guide to individual behaviour and does in fact influence that behaviour. So long as statesmen, and others who influence the conduct of international affairs, agree in thinking that the state has duties, and allow this view to guide their action, the hypothesis remains effective. The acts with which international morality is concerned are performed by individuals not on their own behalf, but on behalf of those fictitious group persons " Great Britain " and " Italy ", and the morality in question is the morality attributed to those " persons ". Any useful examination of international morality must start from recognition of this fact.

Theories of International Morality

Before we consider the moral assumptions which underlie current thinking about international affairs, we must take some account of current theories of international morality. For though it is the assumptions of the ordinary man, not the assumptions of the philosopher, which determine the accepted moral code and govern moral behaviour, the theories of philosophers also exercise an influence on the thought (and, less frequently, on the action) of the ordinary man, and cannot be left altogether out of the picture. Theories of international morality tend to fall into two categories. Realists — and, as we have seen, some who are not realists — hold that relations between states are governed solely by power and that morality plays no part in them. The opposite theory, propounded by most utopian writers, is that the same code of morality is applicable to individuals and to states.

The realist view that no ethical standards are applicable to relations between states can be traced from Machiavelli through Spinoza and Hobbes to Hegel, in whom it found its most finished and thorough-going expression. For Hegel, states are complete and morally self-sufficient entities ; and relations between them express only the concordance or conflict of independent wills not united by any mutual obligation. The converse view that the same standard is applicable to individuals and to states was implicit in the original conception of the personification of the state and has found frequent expression not only in the writings of philosophers, but in the utterances of statesmen of utopian inclinations. " The moral law was not written for men alone in their individual character," said Bright in a speech on foreign policy in 1858, ". . . it was written as well for nations." [1] " We are at the beginning of an age ", said Woodrow Wilson in his address to Congress on the declaration of war in 1917, " in which it will be insisted that the same standards of conduct and of responsibility for wrong shall be observed among nations and their governments that are observed among the individual citizens of civilised states." [2] And when in July 1918 the faithful House tried his hand at the first draft of a League of Nations, Article 1 ran as follows :

> The same standards of honour and ethics shall prevail internationally and in affairs of nations as in other matters. The agreement or promise of a power shall be inviolate.[3]

No corresponding pronouncement was included in the Covenant. But Dr Benes at one of the early Assemblies remarked that the League was " *ipso facto* an attempt to introduce into international relationships the principles and methods employed

---

[1] John Bright, *Speeches on Questions of Public Policy*, p. 479.
[2] *Public Papers of Woodrow Wilson : War and Peace*, i. p. 11.
[3] *Intimate Papers of Colonel House*, ed. C. Seymour, iv p. 28.

. . . in the mutual relations of private individuals ".[1] In his famous Chicago speech of October 5, 1937, President Roosevelt declared that " national morality is as vital as private morality ".[2] But he did not specifically identify them.

Neither the realist view that no moral obligations are binding on states, nor the utopian view that states are subject to the same moral obligations as individuals, corresponds to the assumptions of the ordinary man about international morality. Our task is now to examine these assumptions.

Ordinary Assumptions about International Morality

It is noteworthy that the attempt to deny the relevance of ethical standards to international relations has been made almost exclusively by the philosopher, not by the statesman or the man in the street. Some recognition of an obligation to our fellow-men as such seems implicit in our conception of civilisation ; and the idea of certain obligations automatically incumbent on civilised men has given birth to the idea of similar (though not necessarily identical) obligations incumbent on civilised nations. A state which does not conform to certain standards of behaviour towards its own citizens and, more particularly, towards foreigners will be branded as " uncivilised ". Even Hitler in one of his speeches declined to conclude a pact with Lithuania " because we cannot enter into political treaties with a state which disregards the most primitive laws of human society " ; [3] and he frequently alleged the immorality of Bolshevism as a reason for excluding Soviet Russia from the family of nations. All agree that there is an international moral code binding on states. One of the most important and most clearly recognised items in this code is the obligation not to inflict *unnecessary* death or suffering on other human beings, i.e. death or suffering not necessary for the attainment of some higher purpose which is held, rightly or wrongly, to justify a derogation from the general obligation. This is the foundation of most of the rules of war, the earliest and most developed chapter of international law ; and these rules were generally observed in so far as they did not impede the effective

[1] *League of Nations : Fourth Assembly*, i. p. 144.
[2] *International Conciliation*, No. 334, p. 713.
[3] Speech in the Reichstag, May 21, 1935.

conduct of military operations.[1] A similar humanitarian motive inspired international conventions for the protection of the " backward races " or of national minorities, and for the relief of refugees.

The obligations so far mentioned have been obligations of the state to individuals. But the obligation of state to state is also clearly recognised. The number of synonyms current in international practice for what used to be called " the comity of nations " [2] shews the persistence of the belief that states are members of a comity and have obligations as such. A new state on becoming, in virtue of recognition by other Powers, a member of the international community, is assumed to regard itself as automatically bound, without any express stipulation, by the accepted rules of international law and canons of international morality. As we have seen, the concept of internationalism was so freely used between the two wars for the purpose of justifying the ascendancy of the satisfied Powers that it fell into some disrepute with the dissatisfied Powers. But this natural reaction was not a denial of the existence of an international community so much as a protest against exclusion from the privileges of membership. The result of the Versailles Treaty, wrote Dr. Goebbels, was " to expel Germany from the comity of powerful political countries ", and the function of National Socialism was to " unite the people and once more lead it back to its rightful place in the comity of nations ".[3] During Hitler's visit to Rome in May 1938, Mussolini declared that the common aim of Italy and Germany was " to seek between them and with others a regime of international comity which may restore equally for all more effective guarantees of justice, security and peace ".[4] Constant appeals

---

[1] The rules of war have since 1914 been exposed to an exacting test. The distinction between combatant and non-combatant grows less and less. A deliberate attack on so-called non-combatants may in fact promote important military objectives; and the conception of *unnecessary* suffering, which the belligerent is not entitled to inflict because it is not essential to his military purpose, becomes more and more restricted and difficult to sustain. In short, modern conditions of warfare are doing much to break down, in one important point, a previously existing and effective sense of universal obligation.

[2] Half a dozen synonyms, used quite indiscriminately, are quoted from recent documents by Dr. G. Schwarzenberger (*American Journal of International Law*, xxxiii. p. 59). There is no reason to suspect sarcasm in the reference, in a Japanese Imperial Rescript of 1933, to " the fraternity of nations ".

[3] *Völkischer Beobachter*, April 1, 1939.

[4] *The Times*, May 9, 1938.

were made by both these Powers to the injustice of the conditions imposed on them in the past and the justice of demands now made by them ; and many people in these countries were beyond doubt sincerely and passionately concerned to justify their policy in the light of universal standards of international morality

In particular, the theory that, since states have no moral obligations towards one another, treaties have no binding force, is not held even by those statesmen who exhibit least taste for international co-operation. Every state concludes treaties in the expectation that they will be observed ; and states which violate treaties either deny that they have done so, or else defend the violation by argument designed to shew that it was legally or morally justified. The Soviet Government in the first years of its existence openly violated not only treaties signed by previous Russian governments, but the treaty which it had itself signed at Brest-Litovsk, and propounded a philosophy which seemed to deny international obligation and international morality. But it simultaneously concluded, and offered to conclude, other treaties with the manifest intention of observing them and expecting others to observe them. The German Government accompanied its violation of the Locarno Treaty in 1936 with an offer to enter into a fresh treaty. In neither case is it necessary to doubt the sincerity of the government concerned. Violation of treaties, even when frequently practised, is felt to be something exceptional requiring special justification. The general sense of obligation remains.

The view that the same ethical standard is applicable to the behaviour of states as to that of individuals is, however, just as far from current belief as the view that no standard at all applies to states. The fact is that most people, while believing that states ought to act morally, do not expect of them the same kind of moral behaviour which they expect of themselves and one another.

Many utopian thinkers have been so puzzled by this phenomenon that they have refused to recognise it. Others have sincerely confessed their bewilderment. " Men's morals are paralysed when it comes to international conduct ", observes Professor Dewey ; [1] and Professor Zimmern detects a " rooted prejudice against law and order in the international domain ".[2]

[1] *Foreign Affairs*, March 15, 1923, p. 95.
[2] Zimmern, *Towards a National Policy*, p. 137.

The discrepancy is less surprising than it appears at first sight. Casuists have long been familiar with the problem of incompatibilities between personal, professional and commercial morality. International morality is another category with standards which are in part peculiar to itself. Some of the problems of state morality are common to the whole field of the morality of group persons. Others are peculiar to the state in virtue of its position as the supreme holder of political power. The analogy between the state and other group persons is therefore useful, but not decisive.

Differences between Individual and State Morality

We may now turn to the principal reasons why states are not ordinarily expected to observe the same standards of morality as individuals.

(1) There is the initial difficulty of ascribing to the state, or to any other group person, love, hate, jealousy and other intimate emotions which play a large part in individual morality. It seems plainly incongruous to say, as an eighteenth-century writer said, that " a nation must love other nations as itself ".[1] For this reason, it is sometimes argued that the morality of the state must be confined to that formal kind of morality which can be codified in a set of rules and approximates to law, and that it cannot include such essentially personal qualities as altruism, generosity and compassion, whose obligations can never be precisely and rigidly defined. The state, like a public corporation, can — it is commonly said — be just, but not generous. This does not seem to be entirely true. We have already noted that group persons are commonly assumed to have moral as well as legal rights and obligations. When a bank or a public company subscribes to a Lord Mayor's Fund for assistance to victims of some great disaster, the act of generosity must be attributed not to the directors, whose pockets are not affected, and not to the shareholders, who are neither consulted nor informed, but to the bank or company itself. When the Treasury makes a " compassionate grant " in some case of hardship, the act of compassion is performed not by the official who takes the decision, and not by the

[1] Christian Wolff, quoted in H. Kraus, *Staatsethos*, p. 187.

Chancellor of the Exchequer in his individual capacity, but by the state. Some people expected " the United States " to remit the debts owing to them from European states after the first world war, and criticised their refusal to do so on moral grounds. In other words, paradoxical as it may appear, we do, in certain circumstances, expect states and other group persons, not merely to comply with their formal obligations, but to behave generously and compassionately. And it is precisely this expectation which produces moral behaviour on behalf of a fictitious entity like a bank or a state. Banks subscribe to charitable funds and states make compassionate grants because public opinion expects it of them. The moral impulse may be traced back to individuals. But the moral act is the act of the group person.

Nevertheless, while most people accept the hypothesis that group persons have in certain conditions a moral duty to act altruistically as well as justly, the duty of the group person appears by common consent to be more limited by self-interest than the duty of the individual. In theory, the individual who sacrifices his interests or even his life for the good of others is morally praiseworthy, though this duty might be limited by duty to family or dependents. The group person is not commonly expected to indulge in altruism at the cost of any serious sacrifice of its interests. A bank or public company which failed to pay dividends owing to generous contributions to charities would probably be thought worthy of censure rather than praise. In his presidential campaign of 1932, Franklin Roosevelt referred tauntingly to Mr. Hoover's reputation for humanitarian activities in Europe, and invited him to " turn his eyes from his so-called ' backward and crippled countries ' to the great and stricken markets of Kansas, Nebraska, Iowa, Wisconsin and other agricultural states ".[1] It is not the ordinarily accepted moral duty of a state to lower the standard of living of its citizens by throwing open its frontiers to an unlimited number of foreign refugees, though it may be its duty to admit as large a number as is compatible with the interests of its own people. British supporters of the League of Nations who urged Great Britain to render assistance to victims of " aggression " did not maintain that she should do

[1] Speech at the Metropolitan Opera House, New York, reported in the *New York Times*, November 4, 1932.

this even to the detriment of her vital interests ; they argued that she should render the assistance which she could reasonably afford [1] (just as a bank can reasonably afford to give 500 guineas to the victims of an earthquake). The accepted standard of international morality in regard to the altruistic virtues appears to be that a state should indulge in them in so far as this is not seriously incompatible with its more important interests. The result is that secure and wealthy groups can better afford to behave altruistically than groups which are continually preoccupied with the problem of their own security and solvency ; and this circumstance provides such basis as there is for the assumption commonly made by Englishmen and Americans that the policies of their countries are morally more enlightened than those of other countries.

(2) It is, however, not merely true that the ordinary man does not demand from the group person certain kinds of moral behaviour which are demanded from the individual ; he expects from the group person certain kinds of behaviour which he would definitely regard as immoral in the individual. The group is not only exempt from some of the moral obligations of the individual, but is definitely associated with pugnacity and self-assertion, which become positive virtues of the group person. The individual seeks strength through combination with others in the group ; and his " devotion to his community always means the expression of a transferred egoism as well as of altruism ".[2] If he is strong, he converts the group to the pursuit of his own ends. If he is weak, he finds compensation for his own lack of power to assert himself in the vicarious self-assertion of the group. If we cannot win ourselves, we want our side to win. Loyalty to the group comes to be regarded as a cardinal virtue of the individual, and may require him to condone behaviour by the group person which he would condemn in himself. It becomes a moral duty to promote the welfare, and further the interests, of the group as a whole ; and this duty tends to eclipse duty to a wider community. Acts which would be immoral in the individual may become virtue when performed on behalf of the group person. " If

[1] The League of Nations Union " advocates sanctions only in cases where the number and resources of the governments co-operating on the League's behalf make it reasonably certain that the would-be aggressor will abandon his intention, so that war will not break out at all " (*Headway*, December 1937, p. 232).

[2] R. Niebuhr, *Moral Man and Immoral Society*, p. 40.

we were to do for ourselves what we are doing for Italy," said Cavour to D'Azeglio, " we should be great rogues." [1] The same could truthfully have been said by many directors of public companies and promoters of good causes. " There is an increasing tendency among modern men ", writes Dr. Niebuhr, " to imagine themselves ethical because they have delegated their vices to larger and larger groups." [2] In the same way we delegate our animosities. It is easier for " England " to hate " Germany " than for individual Englishmen to hate individual Germans. It is easier to be anti-Semitic than to hate individual Jews. We condemn such emotions in ourselves as individuals, but indulge them without scruple in our capacity as members of a group.

(3) These considerations apply in some measure to all group persons, though they apply with particular force to the state. There are, however, other respects in which we do not ordinarily demand from the state even the same standard of moral behaviour which we demand from other group persons. The state makes an altogether different kind of emotional appeal to its members from that of any other group person. It covers a far larger field of human activities, and demands from the individual a far more intensive loyalty and far graver sacrifices. The good of the state comes more easily to be regarded as a moral end in itself. If we are asked to die for our country, we must at least be allowed to believe that our country's good is the most important thing in the world. The state thus comes to be regarded as having a right of self-preservation which overrides moral obligation. In the *Cambridge History of British Foreign Policy* published after the war, Professor Holland Rose condones the " discreditable episode " of the seizure of the Danish fleet at Copenhagen in 1807 on the ground of Canning's belief that " the very existence of Great Britain was at stake ".[3] Those who take a different view commonly argue that Canning was mistaken, not that he should have acted otherwise if his belief had been correct.

Other differences between the standards of morality commonly expected of the state and of other group persons arise from the fact that the state is the repository of political power

---

[1] Quoted in E. L. Woodward, *Three Studies in European Conservatism*, p. 297.
[2] R. Niebuhr, *Atlantic Monthly*, 1927, p. 639.
[3] *Cambridge History of British Foreign Policy*, i. pp. 363-4.

and that there is no authority above the state capable of imposing moral behaviour on it, as a certain minimum of moral
behaviour is imposed on other group persons by the state.
One corollary of this is that we are bound to concede to the
state a right of self-help in remedying its just grievances.
Another corollary is the difficulty of securing the observance
by all of a common standard ; for while some moral obligations
are always thought of as absolute, there is a strong tendency
to make the imperativeness of moral obligations dependent on
a reasonable expectation of the performance of the same duty
by others. Conventions play an important part in all morality ;
and the essence of a convention is that it is binding so long
as other people in fact abide by it. Barclays Bank or Imperial
Chemicals Limited would incur moral censure if they employed
secret agents to steal confidential documents from the safes
of rival institutions, since such methods are not habitually
employed by public companies against one another. But no
stigma attaches to " Great Britain " or " Germany " for acting
in this manner ; for such practices are believed to be common
to all the Great Powers, and a state which did not resort to
them might find itself at a disadvantage. Spinoza argued that
states could not be blamed for breaking faith ; for everyone
knew that other states would do likewise if it suited their
interest.[1] One reason why a higher standard of morality is
not expected of states is because states in fact frequently
fail to behave morally and because there are no means of
compelling them to do so.

(4) This brings us to the most fundamental difficulty which
confronts us in our analysis of the moral obligations currently
attributed to the state. It is commonly accepted that the
morality of group persons can only be social morality (a state
or a limited liability company cannot be a saint or a mystic) ;
and social morality implies duty to fellow members of a community, whether that community be a family, a church, a club,
a nation or humanity itself. " No individual can make a
conscience for himself ", writes T. H. Green ; " he always
needs a society to make it for him." [2] In what sense can we
find a basis for international morality by positing a society of
states ?

---

[1] Spinoza, *Tractatus Politicus*, iii. § 14.
[2] T. H. Green, *Prolegomena to Ethics*, p. 351.

Politics, Power and Morality

Is there an International Community?

Those who deny the possibility of an international morality naturally contest the existence of an international community. The English Hegelian Bosanquet, who may be taken as a typical representative of this view, argues that " the nation-state is the widest organisation which has the common experience necessary to found a common life ",[1] and rejects with emphasis " the assumption that humanity is a real corporate being, an object of devotion and a guide to moral duty ".[2] The reply to this would appear to be that a corporate being is never " real " except as a working hypothesis, and that whether a given corporate being is an object of devotion and a guide to moral duty is a question of fact which must be settled by observation and not by theory, and which may be answered differently at different times and places. It has already been shewn that there is in fact a widespread assumption of the existence of a world-wide community of which states are the units and that the conception of the moral obligations of states is closely bound up with this assumption. There is a world community for the reason (and for no other) that people talk, and within certain limits behave, as if there were a world community. There is a world community because, as Señor de Madariaga puts it, " we have smuggled that truth into our store of spiritual thinking without preliminary discussion ".[3]

On the other hand, it would be a dangerous illusion to suppose that this hypothetical world community possesses the unity and coherence of communities of more limited size up to and including the state. If we examine the ways in which the world community falls short of this standard of coherence, we shall have a clue to the underlying reasons for the short-comings of international morality. It falls short mainly in two ways : (i) the principle of equality between members of the community is not applied, and is indeed not easily applicable, in the world community, and (ii) the principle that the good of the whole takes precedence over the good of the part, which is a postulate of any fully integrated community, is not generally accepted.

[1] B. Bosanquet, *The Philosophical Theory of the State*, p. 320.
[2] B. Bosanquet, *Social and International Ideals*, p. 292.
[3] S. de Madariaga, *The World's Design*, p. 3.

Morality in International Politics

The Principle of Equality

(i) The principle of equality within a community is difficult to define. Equality is never absolute, and may perhaps be defined as an absence of discrimination for reasons which are felt to be irrelevant. In Great Britain, the reasons for which some receive higher incomes or pay more taxes than others are (rightly or wrongly) felt to be relevant even by most of those in the less-favoured categories, and the principle of equality is not therefore infringed. But the principle would be infringed, and the community broken, if people with blue eyes were less favourably treated than people with brown, or people from Surrey than people from Hampshire. In many countries, minorities *are* discriminated against on grounds which they feel to be irrelevant, and these minorities cease to feel, and to be regarded, as members of the community.[1]

In the international community such discrimination is endemic. It arises in the first place from the attitude of individuals. Gladstone is said on one occasion to have exhorted an audience of his fellow-countrymen to " remember that the sanctity of life in the villages of the Afghan mountains among the winter snows is no less inviolable in the eyes of the Almighty than your own ".[2] It may safely be said that the eyes of the Almighty are not in this respect those of the great majority of Englishmen. Most men's sense of common interest and obligation is keener in respect of family and friends than in respect of others of their fellow-countrymen, and keener in respect of their fellow-countrymen than of other people. Family and friends form a " face-to-face " group, between whom the sense of moral obligation is most likely to be strong. The members of a modern nation are enabled, through a more or less uniform education, a popular national press, broadcasting and travel facilities, and a skilful use of symbols,[3] to acquire

[1] It is only in recent times that there has begun to be even a presumption that all inhabitants of a territory are members of the community. Like Jews in Nazi Germany, the coloured inhabitants of the Union of South Africa are to-day not regarded as members of the community. In the United States, most white Southerners would hesitate to admit that the negroes are members of the community in the same sense as they are themselves.

[2] Quoted by the Delegate of Haiti in *League of Nations : Fifteenth Assembly*, 6th Committee, p. 43.

[3] " Moral attitudes always develop most sensitively in person-to-person relationships. That is one reason why more inclusive loyalties, naturally more abstract

something of the character of a " face-to-face " group. The ordinary Englishman carries in his mind a generalised picture of the behaviour, daily life, thoughts and interests of other Englishmen, whereas he has no such picture at all of the Greek or the Lithuanian. Moreover, the vividness of his picture of " foreigners " will commonly vary in relation to geographical, racial and linguistic proximity, so that the ordinary Englishman will be likely to feel that he has something, however slight, in common with the German or the Australian and nothing at all in common with the Chinese or the Turk.[1] An American newspaper correspondent in Europe is said to have laid down the rule that an accident was worth reporting if it involved the death of one American, five Englishmen, or ten Europeans. We all apply, consciously or unconsciously, some such standard of relative values. " If it was not that China was so far away," said Neville Chamberlain in the House of Commons on the occasion of Japanese bombing of Chinese cities, " and that the scenes which were taking place there were so remote from our everyday consciousness, the sentiments of pity, horror and indignation which would be aroused by a full observation of those events might drive this people to courses which perhaps they had never yet contemplated." [2] The same *motif* recurred in his national broadcast during the Czecho-Slovak crisis on September 27, 1938 : " How horrible, fantastic, incredible it is that we should be digging trenches and trying on gas-masks here because of a quarrel in a far-away country between people of whom we know nothing ".[3] These words were criticised in many quarters. But there is little doubt that they represented the initial reaction of the ordinary Englishman. Our normal attitude to foreigners is a complete negation of that absence of discrimination on irrelevant grounds which we have

than immediate ones, lose some of their power over the human heart ; and why a shrewd society attempts to restore that power by making a person the symbol of the community " (R. Niebuhr, *Moral Man and Immoral Society*, pp. 52-3).

[1] The variations of feeling are naturally also influenced by current political prejudices.

[2] House of Commons, June 21, 1938 : *Official Report*, col. 936. A correspondent in *The Times*, commenting on " the inconsistencies of compassion " in the international sphere, enquires whether " the world's conscience " regards " 100 dead or destitute Chinese as equivalent to one persecuted Jew ", or whether it is " simply that the Jews are near at hand, while the Chinese are a very long way away, and yellow at that " (*The Times*, November 25, 1938).

[3] N. Chamberlain, *The Struggle for Peace*, p. 275.

recognised as the principle of equality.

This attitude of the individual is reflected in the attitude of states to one another ; and the difficulty is intensified by the structure of the international community. Even if equality between individuals of different countries were recognised, the inequalities between states would be none the less flagrant. The existing inequalities among a handful of known states subject to no external control are infinitely more glaring, more permanent and more difficult to forget than inequalities between the anonymous mass of citizens subject, at any rate in name, to the same law. The importance attached to the idea of equality in international politics is shewn by the number and insistence of the demands based on it. " Most-favoured-nation treatment ", the " Open Door ", " freedom of the seas ", the Japanese claim for the recognition of racial equality in the Covenant of the League of Nations, the old German claim to " a place in the sun ", the more recent German claim to *Gleichberechtigung* or " equality of status ", have all been demands for the application of the principle of equality. The praises of equality were repeatedly sung in the Assemblies and Committees of the League of Nations — mainly, if not exclusively, by delegates of minor Powers.[1] Yet there is little attempt at consistency in the use of the term. Sometimes it merely means formal equality of states before the law. In other contexts, it may mean equality of rights, or equality of opportunity or equality of possessions. Sometimes it seems to mean equality between Great Powers. When Hitler argued that " according to all commonsense, logic and the general principles of high human justice . . . all peoples ought to have an equal share of the goods of the world ",[2] he hardly intended to convey that Lithuania ought to enjoy as much of " the goods of the world " as Germany. Yet if we assume that equality of rights or privileges means proportionate, not absolute, equality, we are

---

[1] Of the Great Powers only France, largely dependent for her position on the support of minor Powers, consistently advocated the principle of equality. " There is not, and we trust there never will be," said M. Blum on one occasion (*League of Nations : Sixteenth Assembly*, Part II, p. 28), " an order of precedence among the Powers forming the international community. Were a hierarchy of States to be established within the League of Nations . . . then the League would be ruined, both morally and materially " — a remarkable statement in view of the hierarchical constitution of the Council.

[2] Speech in the Reichstag of April 28, 1939.

little advanced so long as there is no recognised criterion for determining the proportion. Nor would even this help us much. The trouble is not that Guatemala's rights and privileges are only proportionately, not absolutely, equal to those of the United States, but that such rights and privileges as Guatemala has are enjoyed only by the good-will of the United States. The constant intrusion, or potential intrusion, of power renders almost meaningless any conception of equality between members of the international community.

The Good of the Whole and the Good of the Part

(ii) The other capital shortcoming of the international community is failure to secure general acceptance of the postulate that the good of the whole takes precedence over the good of the part. Great Britain possesses a common national consciousness because the man from Surrey will normally act on the assumption that the good of Great Britain is more important than the good of Surrey. One of the chief obstacles to the growth of a common German national consciousness was the difficulty in persuading Prussians, Saxons and Bavarians to treat the good of Germany as more important than the good of Prussia, Saxony and Bavaria. Now it is clear that, despite pious aspirations, people still hesitate to act on the belief that the good of the world at large is greater than the good of their own country. Loyalty to a world community is not yet powerful enough to create an international morality which will override vital national interests. Yet the conception of a community implies recognition of its good as something which its members are under an obligation to promote, and the conception of morality implies the recognition of principles of a universally binding character. If we refuse altogether to recognise the overriding claims of the whole, can any world community or any kind of international morality be said to exist at all ?

This is the fundamental dilemma of international morality On the one hand, we find the almost universal recognition of an international morality involving a sense of obligation to an international community or to humanity as a whole. On the other hand, we find an almost equally universal reluctance to admit that, in this international community, the good of the part (i.e. our own country) can be less important than the good of the whole. This dilemma is, in practice, resolved in two different ways. The first is the method, which Hitler borrowed from the Darwinian school, of identifying the good of the whole with the good of the fittest. The fittest are by assumption "the bearers of a higher ethic"; [1] and it is only necessary to prove in action that one's country is the fittest in order to establish the identity of its good with the good of the whole. The other method is that of the neo-liberal doctrine of the harmony of interests, of which Woodrow Wilson, Lord Cecil and Professor Toynbee have been quoted as representatives. This doctrine, like every doctrine of a natural harmony of interests, identifies the good of the whole with the security of those in possession. When Woodrow Wilson declared that American principles were the principles of mankind, or Professor Toynbee that the security of the British Empire was "the supreme interest of the whole world",[2] they were in effect making the same claim made by Hitler that their countrymen are "the bearers of a higher ethic"; and the same result is produced of identifying the good of the whole international community with the good of that part of it in which we are particularly interested. Both these methods are equally fatal to any effective conception of international morality.

There is no escape from the fundamental dilemma that every community, and every code of morality, postulates some recognition that the good of the part may have to be sacrificed to the good of the whole. The more explicitly we face this issue in the international community, the nearer we shall be to a solution of our problem. The analogy of the national community, though imperfect, is once more helpful. Modern liberalism, wrote Hobhouse shortly before 1914, "postulates, not that there is an actually existing harmony requiring nothing but prudence and judgment for its effective operation, but only that there is a possible ethical harmony to which . . . men might attain, and that in such attainment lies the social ideal".[3] The word "ethical" betrays the break in the argument. The nineteenth-century "harmony requiring nothing but prudence and judgment for its effective operation" was a harmony of interests. The "ethical harmony" is one achieved by the sacrifice of interests, which is necessary precisely because no

[1] Hitler, *Mein Kampf*, p. 421.   [2] See pp. 77, 79.
[3] L. T. Hobhouse, *Liberalism*, p. 129

natural harmony of interests exists. In the national community, appeals to self-sacrifice are constantly and successfully made, even when the sacrifice asked for is the sacrifice of life. But even in the national community, it would be erroneous to suppose that the so-called " harmony " is established solely through voluntary self-sacrifice. The sacrifice required is frequently a forced one, and the " harmony " is based on the realistic consideration that it is in the " interest " of the individual to sacrifice voluntarily what would otherwise be taken from him by force. Harmony in the national order is achieved by this blend of morality and power.

In the international order, the role of power is greater and that of morality less. When self-sacrifice is attributed to an individual, the sacrifice may or may not be purely voluntary. When self-sacrifice is attributed to a state, the chances are greater that this alleged self-sacrifice will turn out on inspection to be a forced submission to a stronger power. Yet even in international relations, self-sacrifice is not altogether unknown. Many concessions made by Great Britain to the Dominions cannot be explained in terms either of British interests or of submission to the stronger. Concessions made by Great Britain to Germany in the nineteen-twenties, ineffective as they were, were dictated, not wholly by British interests or by fear of Germany's strength, but by a belief in some conception of international morality which was independent of British interests. Any international moral order must rest on some hegemony of power. But this hegemony, like the supremacy of a ruling class within the state, is in itself a challenge to those who do not share it ; and it must, if it is to survive, contain an element of give-and-take, of self-sacrifice on the part of those who have, which will render it tolerable to the other members of the world community. It is through this process of give-and-take, of willingness not to insist on all the prerogatives of power, that morality finds its surest foothold in international — and perhaps also in national — politics. It is, no doubt, useless to begin by expecting far-reaching sacrifices. The standard of what we can reasonably afford must not be pitched too high. But the course most detrimental to international morality is surely to pretend that the German people are the bearers of a higher ethic, or that American principles are the principles of humanity, or that the security of Great Britain is the supreme good of the world, so that no sacrifices at all by one's own nation are in fact necessary. When Professor Zimmern urges "the ordinary man" to "*enlarge* his vision so as to bear in mind that the *public affairs* of the twentieth century are *world affairs*",[1] the most concrete meaning which can be given to this injunction is that the recognition of the principle of self-sacrifice, which is commonly supposed to stop short at the national frontier, should be extended beyond it. It is not certain that ordinary man will remain deaf to such an appeal. If the Chancellor of the Exchequer were to attempt to justify an increase in the income-tax on the ground that it would make us better off, we should dismiss him as a humbug; and this is the kind of argument which is almost invariably used to justify any international policy involving apparent sacrifice of interests. A direct appeal to the need of self-sacrifice for a common good might sometimes prove more effective.

But it is necessary to clear up a further point on which many illusions are current. In the national community, we assume that in this process of self-sacrifice and give-and-take the giving must come principally from those who profit most by the existing order. In the international community, the assumption is commonly made by statesmen and writers of the satisfied Powers that the process of give-and-take operates only within the existing order and that sacrifices should be made by all to maintain that order. International peace, said Mr. Eden once, must be "based on an international order with the nations leagued together to preserve it"; and to this international peace "each nation makes its own contribution because it recognises that therein lies its own enduring interest".[2] The fallacy latent in this and many similar pronouncements is fatal to any workable conception of international morality. The process of give-and-take must apply to challenges to the existing order. Those who profit most by that order can in the long run only hope to maintain it by making sufficient concessions to make it tolerable to those who profit by it least; and the responsibility for seeing that these changes take place as far as possible in an orderly way rests as much on the defenders as on the challengers. This leads us to an examination of the problems of law and change in international politics.

[1] Zimmern, *The Prospects of Civilisation*, p. 26.
[2] Anthony Eden, *Foreign Affairs*, p. 197.

PART FOUR

LAW AND CHANGE

CHAPTER 10

THE FOUNDATIONS OF LAW

No topic has been the subject of more confusion in contemporary thought about international problems than the relationship between politics and law. There is, among many people interested in international affairs, a strong inclination to treat law as something independent of, and ethically superior to, politics. " The moral force of law " is contrasted with the implicitly immoral methods of politics. We are exhorted to establish " the rule of law ", to maintain " international law and order " or to " defend international law " ; and the assumption is made that, by so doing, we shall transfer our differences from the turbulent political atmosphere of self-interest to the purer, serener air of impartial justice. Before adhering to these popular conceptions, we must examine rather carefully the nature and function of law in the international community and its relation to international politics.

The Nature of International Law

International law differs from the municipal law of modern states in being the law of an undeveloped and not fully integrated community. It lacks three institutions which are essential parts of any developed system of municipal law : a judicature, an executive and a legislature.

(1) International law recognises no court competent to give on any issue of law or fact decisions recognised as binding by the community as a whole. It has long been the habit of some states to make special agreements to submit particular disputes to an international court for judicial settlement. The Permanent Court of International Justice, set up under the Covenant of the League, represents an attempt to extend and generalise this habit. But the institution of the Court has not changed

international law : it has merely created certain special obligations for states willing to accept them.

(2) International law has no agents competent to enforce observance of the law. In certain cases, it does indeed recognise the right of an aggrieved party, where a breach of the law has occurred, to take reprisals against the offender. But this is the recognition of a right of self-help, not the enforcement of a penalty by an agent of the law. The measures contemplated in Article 16 of the Covenant of the League, in so far as they can be regarded as punitive and not merely preventive, fall within this category.

(3) Of the two main sources of law — custom and legislation — international law knows only the former, resembling in this respect the law of all primitive communities. To trace the stages by which a certain kind of action or behaviour, from being customary, comes to be recognised as obligatory on all members of the community is the task of the social psychologist rather than of the jurist. But it is by some such process that international law has come into being. In advanced communities, the other source of law — direct legislation — is more prolific, and could not possibly be dispensed with in any modern state. So serious does this lack of international legislation appear that, in the view of some authorities, states do on certain occasions constitute themselves a legislative body, and many multilateral agreements between states are in fact " lawmaking treaties " (*traités-lois*).[1] This view is open to grave objections. A treaty, whatever its scope and content, lacks the essential quality of law : it is not automatically and unconditionally applicable to all members of the community whether they assent to it or not. Attempts have been made from time to time to embody customary international law in multilateral treaties between states. But the value of such attempts has been largely nullified by the fact that no treaty can bind a state which has not accepted it. The Hague Conventions of 1907 on the rules of war are sometimes treated as an example of international legislation. But these conventions were not only not binding on states which were not parties to them, but were not binding on the parties *vis-à-vis* states which were not

[1] The Carnegie Endowment has, for example, given the title *International Legislation* to a collection published under its auspices of " multipartite instruments of general interest ".

parties. The Briand-Kellogg Pact is not, as is sometimes loosely said, a legislative act prohibiting war. It is an agreement between a large number of states " to renounce war as an instrument of national policy in their relations with one another ". International agreements are contracts concluded by states with one another in their capacity as subjects of international law, and not laws created by states in the capacity of international legislators. International legislation does not yet exist.

These shortcomings of international law, serious as they are, do not however deprive it of the title to be considered as law, of which it has all the essential characteristics. In particular, the relation of law to politics will be found to be the same in the international as in the national sphere.

It has been observed that the fundamental question of political philosophy is why men allow themselves to be ruled. The corresponding question which lies at the root of jurisprudence is why men obey the law. Why is law regarded as binding ? The answer cannot be obtained from the law itself any more than a proof of Euclid's postulates can be obtained from Euclid. Law proceeds on the assumption that the question has been satisfactorily disposed of. But it is a question which cannot be burked by those who seek to justify the " rule of law ". It applies to international as well as to municipal law. In international law, it sometimes takes the form of the question whether, and on what grounds, treaties are binding. The legal answer to this question is that treaties are binding in international law, which includes the rule (subject to some reservations which will be discussed presently) that treaties must be kept. But what the questioner probably means to ask is : Why is international law, and with it the rule that treaties must be kept, binding, and should they be regarded as binding at all ? These are not questions which can be answered by international law. It is the purpose of this chapter to enquire in what domain the answer to them should be sought, and what that answer should be.

In approaching the problem of the ultimate authority of law, we shall find the same fundamental divergence which we have traced in the field of politics between utopians, who think in terms of ethics, and realists, who think in terms of power. Among students of law, the utopians are commonly known as " naturalists ", who find the authority of law in natural law, and the realists as "positivists", who find the authority of law in the will of states. The terminology tends to become blurred and fluctuating. Some utopians purport to reject natural law, and adopt some other standard such as reason, utility, "objective right",[1] "ultimate sense of right",[2] or a "fundamental norm". Conversely, some positivists such as Spinoza purport to accept natural law, but empty it of its meaning by virtually identifying it with the right of the stronger. Other positivists fly the colours of "the historical school of law" or of "the economic interpretation of law". But the fundamental divergence remains between those who regard law primarily as a branch of ethics, and those who regard it primarily as a vehicle of power.

The Naturalist View of Law

The naturalist view of law, like the utopian view of politics, has a longer history behind it than the positivist or realist view. In primitive communities, law is bound up with religion and, until a fairly late stage of human development, always appears to emanate from a god or a divinely appointed law-giver. The secular civilisation of the Greeks divorced law from religion, but not from morality. Greek thinkers found in the conception of natural law a higher unwritten law from which man-made law derived its validity and by which it could be tested. The acceptance of Christianity by the Roman Empire restored divine authority. Natural law was for a time identified with divine law ; and it was only at the Renaissance that it resumed its independent role as a non-theological ethical standard. As we have seen, the seventeenth and eighteenth centuries revived in a new form the identification of natural law with reason. "Law in general", says Montesquieu, "is human reason, inasmuch as it governs all the peoples of the earth."[3] It was under these auspices that modern international law was created by Grotius and his successors to meet the needs of the new nation-states which had arisen on the ruins of the mediaeval world. International law was therefore by origin strongly utopian. This was necessary and inevitable. The new con-

[1] Duguit, *Traité de droit constitutionnel*, i. p. 16.
[2] Krabbe, *The Modern Idea of the State* (Engl. transl.), p. 110.
[3] Montesquieu, *Esprit des Lois*, Book I. ch. iii.

ventions which came more or less effectively to govern relations between states grew no doubt out of practical needs. But they could never have secured as wide an acceptance as they did if they had not been treated as binding in virtue of natural law and universal reason. But here we shall note the recurrence of a paradox which is also apparent in the political field. Where practice is least ethical, theory becomes most utopian. Owing to the more primitive state of development of the international community, morality plays a smaller effective role in the practice of international law than of municipal law. In theories of international law, utopia tends to predominate over reality to an extent unparalleled in other branches of jurisprudence. Moreover, this tendency is greatest at periods when anarchy is most prevalent in the practice of nations. During the nineteenth century, a comparatively orderly period in international affairs, international jurisprudence took on a realist complexion. Since 1919, natural law has resumed its sway, and theories of international law have become more markedly utopian than at any previous time.

The modern view of natural law differs, however, in one important respect from the view which prevailed down to the end of the eighteenth century. Prior to that time, natural law had always been conceived as something essentially static, a fixed and eternal standard of right which must, in the nature of things, be the same yesterday, to-day and for ever. The historical tendency of nineteenth-century thought, which at first threatened to eclipse natural law altogether, gave it a new direction; and towards the end of the century there emerged the new conception of " natural law with a variable content ".[1] Natural law, in this interpretation, connotes no longer something external, fixed and invariable, but men's innate feeling at any given time or place for what " just law " ought to be. This revised definition of natural law helps us a little. It gets over the old crux that slavery was at one time thought to be sanctioned, and at another time to be prohibited, by natural law, or that private property is in some places regarded as a natural right and in other places as an infringement of natural right. We are now asked to treat law as binding because it is an emanation not of some eternal ethical

[1] The phrase comes from Stammler, whose *Lehre von dem richtigen Rechte* (1902–7) has been translated into English under the title *The Theory of Justice.*

principle, but of the ethical principles of a given time and community. This is, at any rate, a part of the truth. The ethical character of the impulse which lies behind many rules of law, municipal and international, including the rule of international law that treaties should be kept, will not be denied by any reasonable person. The prevalence in most European languages of words which bestride the frontier between law and ethics betrays a widespread conviction of the close relationship between them.

Nevertheless, this explanation why law is regarded as binding will turn out, on further examination, to be inadequate and in some degree misleading. The main crux about natural law is not that people differ from time to time and from place to place about what particular rules it prescribes (for this crux might be surmounted by the "variable" theory), but that natural law (or reason or "objective right" or any of its other substitutes) can be just as easily invoked to incite disobedience to the law as to justify obedience to it. Natural law has always had two aspects and two uses. It can be invoked by conservatives to justify the existing order, as when the rights of rulers or the rights of property are alleged to rest on natural law. It can equally be invoked by revolutionaries to justify rebellion against the existing order. There is in natural law an anarchic element which is the direct antithesis of law. Theories of law which seek the ultimate authority of law in its ethical content can explain only why good laws (or laws regarded as good at a given time and place) are regarded as binding. Yet there is a fairly general consensus of opinion which regards as binding even laws recognised as bad ; and it may be doubted whether any community could long survive in which such an opinion did not prevail. It is commonly admitted that there *may* be a right or duty to disobey a bad law. But in such cases, a conflict is recognised to exist between two duties ; and it is generally felt that only the most exceptional circumstances justify a decision in favour of the duty to disobey. No theory of law seems adequate which explains that law is regarded as binding because it conforms to natural law or because it is good.

Law and Change

The Realist View of Law

The positivist or realist view of law was first clearly and explicitly stated by Hobbes, who defined law as a command : *Ius est quod iussum est.* Law is thus divorced altogether from ethics. It may be oppressive or otherwise immoral. It is regarded as binding because there is an authority which enforces obedience to it. It is an expression of the will of the state, and is used by those who control the state as an instrument of coercion against those who oppose their power. The law is therefore the weapon of the stronger. That contradictory thinker Rousseau, who elsewhere treats law as the antithesis of despotism, has recorded this view in emphatic terms : " The spirit of the laws of all countries is always to favour the strong against the weak and him that has against him that has not. This drawback is inevitable, and there are no exceptions to it." [1] According to Marx, all law is a " law of inequality ".[2] The principal contribution of Marxism to the problem is its insistence on the relativity of law. Law reflects not any fixed ethical standard, but the policy and interests of the dominant group in a given state at a given period. Law, as Lenin puts it, is " the formulation, the registration of power relations " and " an expression of the will of the ruling class ".[3] The realist view of the ultimate basis of law is well summed up by Professor Laski : " Legal rules are always seeking to accomplish an end deemed desirable by some group of men, and it is only by constant formulation of what that end is that we can obtain a realistic jurisprudence ".[4]

The realist answer to the question why law is regarded as binding contains, like the " naturalist " answer, a part of the truth. Some people do in fact obey some laws because law-breaking will bring them into unwelcome contact with the police and the courts. But no community could survive if most of its members were law-abiding only through an ever-present fear of punishment. As Laud says, " no laws can be binding if there be no conscience to them " ; [5] and there is plenty of evidence of the difficulty of enforcing laws which

[1] Rousseau, *Émile*, Book IV.
[2] Marx and Engels, *Works* (Russian ed.), xv. p. 272.
[3] Lenin, *Works* (2nd Russian ed.), xv. p. 330 ; xii. p. 288.
[4] *Representative Opinions of Mr. Justice Holmes*, ed. Laski, Introduction.
[5] Laud, Sermon IV, *Works*, i. p. 112.

seriously offend the conscience of the community or of any considerable part of it. Law is regarded as binding because it represents the sense of right of the community : it is an instrument of the common good. Law is regarded as binding because it is enforced by the strong arm of authority : it can be, and often is, oppressive. Both these answers are true ; and both of them are only half-truths.

Law as a Function of Political Society

If then we wish to reconcile these contradictory and inadequate half-truths, and to find a single answer to the question why law is regarded as binding, we must seek it in the relationship of law to politics. Law is regarded as binding because, if it were not, political society could not exist and there could be no law. Law is not an abstraction. It " can only exist within a social framework. . . Where there is law, there must be a society within which it is operative." [1] We need not dwell on the old controversy whether, as the positivists held, the state creates law, or as the naturalists held, law creates the state. It is sufficient to say that no political society can exist without law, and that law cannot exist except in a political society.[2] The point has been clearly put by a contemporary German writer :

All law is always the expression of a community. Every legal community (*Rechtsgemeinschaft*) has a common view of law (*Recht*) determined by its content. It is an impossible undertaking to seek to construct a legal community without such a common view, or to establish a legal community before a minimum common view about the content of the community's law has been attained.[3]

Politics and law are indissolubly intertwined ; for the relations of man to man in society which are the subject-matter of the

[1] Zimmern, *International Affairs*, xvii. (January–February 1938), p. 12.

[2] " We shall no longer ask whether the state is prior to law, or law is prior to the state. We shall regard them both as inherent functions of the common life which is inseparable from the idea of man. They will both be primordial facts : they will both have been given, as seeds or germs, coevally with man himself : they will both appear, as developed fruits, simultaneously with one another and in virtue of one another " (Gierke, *Natural Law and the Theory of Society*, Engl. transl., p. 224).

[3] F. Berber, *Sicherheit und Gerechtigkeit*, p. 145.

one are the subject-matter of the other. Law, like politics, is a meeting place for ethics and power.

The same is true of international law, which can have no existence except in so far as there is an international community which, on the basis of a " minimum common view ", recognises it as binding. International law is a function of the political community of nations. Its defects are due, not to any technical shortcomings, but to the embryonic character of the community in which it functions. Just as international morality is weaker than national morality, so international law is necessarily weaker and poorer in content than the municipal law of a highly organised modern state. The tiny number of states forming the international community creates the same special problem in law as in ethics. The evolution of general rules equally applicable to all, which is the basis of the ethical element in law, becomes extremely difficult. Rules, however general in form, will be constantly found to be aimed at a particular state or group of states ; and for this reason, if for no other, the power element is more predominant and more obvious in international than in municipal law, whose subjects are a large body of anonymous individuals. The same consideration makes international law more frankly political than other branches of law.

Once therefore it is understood that law is a function of a given political order, whose existence alone can make it binding, we can see the fallacy of the personification of law implicit in such popular phrases as "the rule of law" or "the government of laws and not of men". The man in the street tends to personify law as something which, whether he approves it or not, he recognises as binding on him ; and this personification is as natural for everyday purposes as the personification of the state. It is, nevertheless, dangerous to clear thinking. Law cannot be self-contained ; for the obligation to obey it must always rest on something outside itself. It is neither self-creating nor self-applying. " There are men who govern ", says a Chinese philosopher, " but there are no laws that govern." [1] When Hegel finds the embodiment of the highest

---

[1] Hsun-tze, quoted in Liang Chi-chao, *History of Chinese Political Thought*, p. 137. A perfect illustration of the confusion which results from treating law as something self-contained and self-applying may be found in a reported dictum of Mr. Winston Churchill: " There must be the assurance that some august international tribunal shall be established which will uphold, enforce and itself obey

moral good in the state, we are entitled to ask, What state ? or, better, Whose state ? When modern writers on international politics find the highest moral good in the rule of law, we are equally entitled to ask, What law ? and Whose law ? The law is not an abstraction. It cannot be understood independently of the political foundation on which it rests and of the political interests which it serves.

We shall also have no difficulty in detecting the fallacy in the common illusion that law is more moral than politics. A transaction, by becoming legal, does not become moral. To pay a workman less than a living wage is not any more moral because the wage is fixed in a contract signed by the workman and valid in law. The annexations of French territory by Germany in 1871 and of German territory by the Allies in 1919 may have been moral or immoral. But they are not made any more moral by the fact that they were registered in treaties signed by the defeated Powers and valid in international law. It is not in itself any more moral to deprive Jews of their property by a law to that effect than simply to send storm-troopers to evict them. The laws of the Medes and Persians were probably not conspicuously moral. If the law is " always seeking to accomplish an end deemed desirable by some group of men," the ethical character of the law is obviously conditioned by that end. Political action can be, and often is, invoked to remedy immoral or oppressive law. The peculiar quality of law which makes it a necessity in every political society resides not in its subject-matter, nor in its ethical content, but in its stability. Law gives to society that element of fixity and regularity and continuity without which no coherent life is possible. It is the fundamental basis of organised political society that the rights and duties of citizens in relation both to one another and to the state should be defined by law. Law which is uncertain in its interpretation or capricious in its application fails to fulfil its essential function.

Stability and continuity are, however, not the only requisites of political life. Society cannot live by law alone, and law cannot be the supreme authority. The political arena is the

the law " (*Manchester Guardian*, December 12, 1938). If Mr. Churchill had paused to ask *who* would establish the august tribunal, *who* would enforce its decisions, *who* would make the law and *who* would see that the tribunal obeyed it, the implications of this apparently simple proposition would have become apparent.

scene of a more or less constant struggle between conservatives, who in a general way desire to maintain the existing legal situation, and radicals, who desire to change it in important respects ; and conservatives, national and international, have the habit of posing as defenders of the law and of decrying their opponents as assailants of it. In democracies, this struggle between conservatives and radicals is carried on openly in accordance with legal rules. But these rules are themselves the product of a pre-legal political agreement. Every system of law presupposes an initial political decision, whether explicit or implied, whether achieved by voting or by bargaining or by force, as to the authority entitled to make and unmake law. Behind all law there is this necessary political background. The ultimate authority of law derives from politics.

CHAPTER 11

THE SANCTITY OF TREATIES

ONE of the functions of law necessary to civilised life is to protect rights which have been created by private contracts concluded in a manner recognised by the law as valid. International law upholds, with some reservations, rights created by international treaties and agreements. This principle is essential to the existence of any kind of international community and is, as we have seen, recognised in theory by all states. The fact that the only written obligations of states are those contained in treaties, and that customary international law is limited in scope and sometimes uncertain in content, has given to treaties a more prominent place in international law than is occupied by contracts in municipal law. Indeed the contents of treaties are sometimes misleadingly spoken of as if they were a part of international law itself, though nobody would regard the provisions of a contract between Smith and Robinson as a part of municipal law. The principle of the sanctity of treaties has thus been thrown into undue relief, which was further intensified by the controversy over the peace treaties of 1919–20. Between the two wars writers, especially those from countries interested in the maintenance of the peace settlement, attempted to treat the rule *pacta sunt servanda* not merely as a fundamental rule of international law, but as the corner-stone of international society — an attitude mockingly described by a German writer as " *pacta-sunt-servanda*-ism ".[1] The issue has become one of the most contentious in the whole field of international politics ; and confusion has often been caused by failure to distinguish between " the sanctity of treaties " as a rule of international law and " the sanctity of treaties " as a principle of international ethics.

[1] Walz in *Deutsches Recht*, Jg. IV. (1934), p. 525. Professor Lauterpacht's remark that the rule *pacta sunt servanda* " constitutes the highest, irreducible, final criterion " in international society (*The Function of Law in the International Community*. p. 418) is a good example of the attitude criticised.

Law and Change

The Legal and Moral Validity of Treaties

In spite of the universal recognition by all countries that treaties are in principle legally binding, international law before 1914 was reluctant to treat as absolute the binding character of treaty obligations. Account had to be taken of the fact that, while states interested in the maintenance of the *status quo* vigorously asserted the unconditional validity of treaties in international law, a state whose interests were adversely affected by a treaty commonly repudiated it as soon as it could do so with impunity. France in 1848 announced that " the treaties of 1815 are no longer valid in the eyes of the French Republic ".[1] Russia in 1871 repudiated the Straits Convention placing restrictions on the passage of her warships which had been imposed on her at the conclusion of the Crimean War. These were merely the most conspicuous of several similar nineteenth-century occurrences. To meet such conditions, international lawyers evolved the doctrine that a so-called *clausula rebus sic stantibus* was implicit in every treaty, i.e. that the obligations of a treaty were binding in international law so long as the conditions prevailing at the time of the conclusion of the treaty continued, and no longer. This doctrine, if carried to its logical conclusion, would appear to lead to the position that a treaty has no authority other than the power relationship of the parties to it, and that when this relationship alters the treaty lapses. This position was not infrequently adopted. " Every treaty ", wrote Bismarck in a famous phrase, " has the significance only of a constatation of a definite position in European affairs. The reserve *rebus sic stantibus* is always silently understood." [2] The same effect is produced by the doctrine occasionally propounded that a state enjoys the unconditional right to denounce any treaty at any time. This view was stated in its most uncompromising form by Theodore Roosevelt : " The nation has as a matter of course a right to abrogate a treaty in a solemn and official manner for what she regards as a sufficient cause, just exactly as she has a right to declare war or exercise another power for a sufficient cause ".[3] Woodrow Wilson observed in private conversation during the Peace Conference

[1] Lamartine's Circular of March 5, 1848, published in the *Moniteur* of that date.
[2] Bismarck, *Gedanken und Erinnerungen*, ii. p. 258.
[3] Quoted in H. F. Pringle, *Theodore Roosevelt*, p. 309.

that, when he was a teacher of international law, he had always supposed that a state had the power to denounce any treaty by which it was bound at any time.[1] In 1915, a distinguished neutral international lawyer of the " naturalist " school wrote of the rule *pacta sunt servanda* that " nobody regards it as a rule of law which is valid without exception either within or without the state ".[2]

Even Great Britain which, as the strongest Power in the world, had most interest in upholding the validity of treaties, was manifestly disinclined to accept the view that treaty obligations were unconditionally binding. The most famous example is that of the Belgian Guarantee Treaty of 1839, under which the principal European Powers, including Great Britain, bound themselves jointly and severally to resist any violation of the neutrality of Belgium by one of their number. In 1870 Gladstone told the House of Commons, in a passage which was cited with approval by Grey in his speech of August 3, 1914, that he was " not able to subscribe to the doctrine of those who have held in this House what plainly amounts to an assertion that the simple fact of the existence of the guarantee is binding on every party of it, irrespective altogether of the particular position in which it may find itself at the time that the occasion for acting on the guarantee arises ". Such an interpretation Gladstone thought " rigid " and " impracticable ".[3] A confidential minute written in 1908 by Lord Hardinge, then Permanent Under-Secretary of State for Foreign Affairs, was conceived in the same spirit :

> The liability undoubtedly exists . . . but whether we could be called on to carry out our obligation and to vindicate the neutrality of Belgium in opposing its violation must necessarily depend upon our policy at the time and the circumstances of the moment. Supposing that France violated the neutrality of Belgium in a war against Germany it is, under present circumstances, doubtful whether England or Russia would move a finger to maintain Belgian neutrality, while, if the neutrality of Belgium were violated by Germany, it is probable that the converse would be the case.

[1] Miller, *The Drafting of the Covenant*, i. p. 293.
[2] Krabbe, *The Modern Idea of the State* (Engl. transl.), p. 266.
[3] Quoted in Grey, *Speeches on Foreign Affairs, 1904-1914*, p. 307.

Grey, commenting in a further minute, merely observed that
this reflexion was "to the point ".[1]

Another principle not less elastic than the *clausula* has
sometimes been invoked to justify non-fulfilment of international
obligations — the principle of "necessity" or "vital interests ".
It is a well-known legal maxim that nobody can be called on
to perform the impossible ; and the impossible is sometimes
held in international law to include acts detrimental to the vital
interests (meaning primarily the security) of the state. Some
writers have specifically held that every state has a legal right
of self-preservation which overrides any obligation to other
states. This view is likely to carry particular weight in time of
war. In its note of protest against British blockade measures
in December 1914, the United States Government laid it down
as the principle of international law that belligerents should
not interfere with neutral commerce "unless such interference
is manifestly an imperative necessity to protect their national
safety, and then only to the extent that it is a necessity ". The
British Government gratefully accepted this interpretation, and
was thenceforth able to justify its blockade activities on the
uncontested ground of an "imperative necessity" whose
requirements nobody was as well qualified as itself to assess.[2]
In such emergencies, the layman is apt to discard legal niceties
and arrive at the same result by other methods. At the time
of the Jameson Raid, *The Times* published a poem by the
Poet Laureate which opened with these disarming lines :

> Let lawyers and statesmen addle
> Their pates over points of law :
> If sound be our sword and saddle
> And gun-gear, who cares one straw ? [3]

"Damn the law, I want the Canal built " was a saying popularly attributed to Theodore Roosevelt at the time of the
Panama crisis. In 1939 a Japanese "naval spokesman ",
commenting on the boarding of foreign vessels in Chinese
waters by Japanese patrols, is reported to have said : " It is
not a question of having the right to do this. It is something

---

[1] *British Documents on the Origin of the War*, ed. Gooch and Temperley,
viii. pp. 377-8.
[2] The correspondence was published in Cmd. 7816 of 1915.
[3] *The Times*, January 11, 1896.

which is necessary and we are doing it." [1] " Once it [i.e. the nation] is in danger of oppression or annihilation," wrote Hitler, " the question of legality plays a subordinate role." [2]

Indeed, where justification is explicitly or implicitly offered for the non-fulfilment of treaty obligations, it is often difficult to discover from the words used whether the alleged justification is based on legal or on moral grounds. Is the view taken that, by the operation of *clausula rebus sic stantibus* or for some other reason, the obligation is no longer binding in law ? Or is the legal obligation admitted, and is it argued that the state is entitled to disregard the law on the ground that it is immoral, unreasonable or impracticable, just as the citizen is sometimes morally entitled to disregard the national law ? Broadly speaking, it may be said that prior to 1914 the rule *pacta sunt servanda* was elastically interpreted and the non-fulfilment of obligations was apt to be defended as legally admissible, whereas since 1919 the interpretation of the rule has tended to become more rigid, and non-fulfilment has been defended mainly on the ground that considerations of reason or morality entitled the state to disregard its strictly legal obligation. The dilemma of international law is that of ecclesiastical dogma. Elastic interpretation adapted to diverse needs increases the number of the faithful. Rigid interpretation, though theoretically desirable, provokes secessions from the church. It cannot be doubted that the more frequent and open repudiation of the rules of international law since 1919 has been due in part to the well-intentioned efforts of the victorious Powers to strengthen those rules and to interpret them with greater rigidity and precision.

An examination of the numerous breaches of treaty obligations during this period yields less definite results than might have been expected ; for the state concerned in many cases defended itself either by denying that any breach of treaty obligations has occurred, or by alleging that the treaty had in the first instance been violated by the other party. In December 1932, the French Chamber of Deputies refused to carry out the French War debt agreement with the United States on the ground that " the determining circumstances " had changed since the conclusion of the agreement six years earlier — the nearest approach since 1919 to an explicit invoca-

[1] *The Times*, May 26, 1939. [2] Hitler, *Mein Kampf*, p. 104.

tion of the *clausula rebus sic stantibus*.[1]   The British default on the Anglo-American War debt agreement was justified on the ground of " economic necessity ".   But the main ground of the argument was not legal, but moral : the burden imposed by the agreement was " unreasonable " and " inequitable ".[2] *The Times* took the view that the debt " had not the same moral validity as an ordinary commercial transaction ".[3]   At an earlier stage, Neville Chamberlain, then Chancellor of the Exchequer, had explicitly admitted that the obligation was legally binding, but had appealed to other obligations which might be rated higher than those of law :

> When we are told that contracts must be kept sacred, and that we must on no account depart from the obligations which we have undertaken, it must not be forgotten that we have other obligations and responsibilities, obligations not only to our own countrymen, but to many millions of human beings throughout the world, whose happiness or misery may depend upon how far the fulfilment of these obligations is insisted upon on the one side and met on the other.[4]

In repudiating the military clauses of the Versailles Treaty in March 1935, Germany based her action on the alleged failure of the other parties to the treaty to implement their own obligations to disarm.   A year later, the repudiation of the Locarno Treaty was justified on the ground that, through the action of France in concluding the Franco-Soviet Pact, the treaty had " ceased in practice to exist ".[5]   These were at any rate ostensibly legal arguments.   But in a public speech shortly after the occupation of the Rhineland, Hitler rejected the legal in favour of the moral plea : " If the rest of the world clings to the letter of treaties, I cling to an eternal morality ".[6]

---

[1] Resolution of December 14, 1932, in *Documents on International Affairs, 1932*, pp. 80-82.

[2] The quotations are from the British note of June 4, 1934 (Cmd. 4609).

[3] *The Times*, June 2, 1934.

[4] Speech in the House of Commons, December 14, 1932, in *Documents on International Affairs, 1932*, p. 128.

[5] *Diplomatic Discussions Directed Towards a European Settlement*, Cmd. 5143, p. 78.

[6] Quoted in Toynbee, *Survey of International Affairs, 1936*, p. 319.   Such pleas are not peculiarly modern and have often been regarded as legitimate.   As recently as 1908, a distinguished English historian used of Pitt words which, with the bare change of proper names, are precisely apposite to Hitler's attitude : " His

The Sanctity of Treaties

On the whole, therefore, it may be said that breaches of treaties between the two wars were excused, not on the legal ground of derogations admitted by international law to the principle of the sanctity of treaties, but on the ethical ground that certain treaties, though legally binding, lack moral validity. It was not denied that breaches of such treaties are technical breaches of international law ; but they were condoned on the ground that the treaties themselves were an offence against international morality.   It is important for the student of international ethics and international law to study the qualities which were popularly supposed to make treaties morally disreputable and therefore morally invalid.

Treaties Signed Under Duress

In the first place, it came to be felt that there was a moral taint about treaties signed under duress. This feeling attached itself mainly to the Versailles Treaty, signed by Germany under the duress of a five-day ultimatum. German propaganda worked hard to popularise the conception of the Versailles Treaty as a *Diktat* which had no moral validity; and the idea enjoyed widespread currency after the conclusion of the Locarno Treaty, when British and French statesmen rashly vied with Stresemann in emphasising the moral significance of the voluntary acceptance by Germany of some of the obligations accepted under duress at Versailles. The attitude adopted to treaties concluded under duress is dependent on the attitude adopted to war ; for every treaty which brings a war to an end is almost inevitably accepted by the loser under duress. So long therefore as any kind of war whatever is recognised as moral, treaties concluded under duress cannot be unconditionally condemned as immoral. The moral objections most frequently expressed against the Versailles Treaty seem, in

support of the British claim as ' from God and Nature ' to override the artificial restrictions of unjust treaties, his denunciation of the Convention of the Pardo as ' a stipulation for the national ignominy ', voiced the inarticulate sentiment of the new England " (*Quarterly Review*, October 1908, p. 325). A later passage in the same article runs as follows : " By the alchemy of his own intense vision and political ideals, he imposed on England a conception of national development and national ends based on an ideal of Imperialist expansion to realise which the nation must sacrifice everything or cease to believe in its own right and power to exist " (*ibid*. pp. 334-5). It is interesting to observe that the writer clearly regarded these phrases as eulogistic.

fact, to have been based not so much on its signature under duress as on the severity of its contents, and on the fact that the Allied Governments, reversing the procedure followed at all important peace conferences down to and including that of Brest-Litovsk, refused to engage in oral negotiations with the plenipotentiaries of the defeated Power. This act of unwisdom probably discredited the treaty more than the ultimatum which preceded its signature

Inequitable Treaties

Secondly, the view was commonly taken that treaties may be morally invalidated by the character of their contents. There cannot indeed be any rule of international *law* corresponding to the rule of municipal law voiding contracts which are "immoral" or "contrary to public policy". The absence of an international political order makes impossible any *legal* definition of international public policy or of what is internationally immoral.[1] But those who regard the contents of a given international treaty as immoral will, generally speaking, concede to the injured state the moral right to repudiate it ; for international law provides no other means of redress. It should, moreover, be observed that there is a tendency to concede the same moral right to repudiate a treaty which is not, properly speaking, immoral, but which is inequitable in the sense that it imposes conditions flagrantly incompatible with the existing relations of power between the contracting parties. The disarmament clauses of the Versailles Treaty were widely regarded as lacking in validity because it was unreasonable to impose a position of permanent inferiority on a Great Power. In general, the reproach was levelled against the Versailles Treaty that it sought to perpetuate the temporary weakness of Germany due to her collapse at the end of the War. This argument is

---

[1] Some German writers after 1919 tried to maintain that treaties are invalid in international law if they conflict with the "natural law of nations". The literature is reviewed by Verdross, *American Journal of International Law*, xxxi. (October 1937), pp. 571 *sqq.* But this view has found little support elsewhere. On the occasion of a judgment by the Permanent Court of International Justice in 1934. the German judge, in an individual opinion, expressed the view that the Court "would never apply a convention whose contents were contrary to *bonnes mœurs*" (*Permanent Court of International Justice*, Series A/B No. 63, p. 150). But the Court as such never appears to have committed itself to this proposition.

not perhaps strictly ethical, since it is rooted in the power position and recognises a moral right based simply on strength. But it is an illustration of the curious way in which power and ethics are intertwined in all political problems. A somewhat similar case arose in connexion with Article 16 of the Covenant of the League of nations. When the United States failed to ratify the Covenant, it was widely felt that the obligations imposed by that Article were no longer morally binding, since members of the League could not reasonably be expected to take measures which might bring on them the enmity of so powerful a country. The test of what is commonly recognised as reasonable applies to the moral validity of treaties as to other problems of international morality.

Treaties as Instruments of Power

The third consideration which is sometimes invoked to deny the morally binding character of international treaties is of a more sweeping kind. It is designed to cast doubts on the moral credit not of particular treaties, but of all treaties as being by their nature instruments of power and therefore devoid of moral value. A Marxist writer has argued that, in capitalist society, the legal enforcement of contracts is merely a method of using the power of the state to protect and further the interests of the ruling class.[1] In the same way, it can be maintained with considerable show of reason that insistence on the legal validity of international treaties is a weapon used by the ruling nations to maintain their supremacy over weaker nations on whom the treaties have been imposed. Such an argument is implicit in the realist view of law as an oppressive instrument of power divorced from ethics.

The argument is assisted by the elastic and inconsistent manner in which the doctrine of the sanctity of treaties has been applied in the practice of states. In 1932–33, the French and British Governments were insisting with particular vehemence that the disarmament clauses of the Versailles Treaty were legally binding on Germany, and could be revised only with the consent of the interested Powers. In December 1932, the French Chamber of Deputies found reasons for refusing to carry out the French war debt agreement with the United

---

[1] Renner, *Die Rechtsinstitute des Privatrechts und ihre soziale Funktion*, p. 55.

States. In June 1933, the British Government ceased to pay the regular instalments due under its war debt agreement, substituting minor "token payments"; and a year later these token payments came to an end. Yet in 1935 Great Britain and France once more joined in a solemn condemnation of Germany for unilaterally repudiating her obligations under the disarmament clauses of the Versailles Treaty. Such inconsistencies are so common that the realist finds little difficulty in reducing them to a simple rule. The element of power is inherent in every political treaty. The contents of such a treaty reflect in some degree the relative strength of the contracting parties. Stronger states will insist on the sanctity of the treaties concluded by them with weaker states. Weaker states will renounce treaties concluded by them with stronger states so soon as the power position alters and the weaker state feels itself strong enough to reject or modify the obligation. Since 1918, the United States have concluded no treaty with a stronger state, and have therefore unreservedly upheld the sanctity of treaties. Great Britain concluded the war debt agreement with a country financially stronger than herself, and defaulted. She concluded no other important treaty with a stronger Power and, with this single exception, upheld the sanctity of treaties. The countries which had concluded the largest number of treaties with states stronger than themselves, and subsequently strengthened their position, were Germany, Italy and Japan; and these are the countries which renounced or violated the largest number of treaties. But it would be rash to assume any *moral* distinction between these different attitudes. There is no reason to assume that these countries would insist any less strongly than Great Britain or the United States on the sanctity of treaties favourable to themselves concluded by them with weaker states.

The case is convincing as far as it goes. The rule *pacta sunt servanda* is not a moral principle, and its application cannot always be justified on ethical grounds. It is a rule of international law; and as such it not only is, but is universally recognised to be, necessary to the existence of an international society. But law does not purport to solve every political problem; and where it fails, the fault often lies with those who seek to put it to uses for which it was never intended. It is no reproach to law to describe it as a bulwark of the existing order.

The essence of law is to promote stability and maintain the existing framework of society ; and it is perfectly natural everywhere for conservatives to describe themselves as the party of law and order, and to denounce radicals as disturbers of the peace and enemies of the law. The history of every society reveals a strong tendency on the part of those who want important changes in the existing order to commit acts which are illegal or which can plausibly be denounced as such by conservatives. It is true that in highly organised societies, where legally constituted machinery exists for bringing about changes in the law, this tendency to illegal action is mitigated. But it is never removed altogether. Radicals are always more likely than conservatives to come into conflict with the law.

Before 1914, international law did not condemn as illegal resort to war for the purpose of changing the existing international order ; and no legally constituted machinery existed for bringing about changes in any other way. After 1918 opinion condemning " aggressive " war became almost universal, and nearly all the nations of the world signed a pact renouncing resort to war as an instrument of policy. While therefore resort to war for the purpose of altering the *status quo* now usually involves the breach of a treaty obligation and is accordingly illegal in international law, no effective international machinery has been constituted for bringing about changes by pacific means. The rude nineteenth-century system, or lack of system, was logical in recognising as legal the one effective method of changing the *status quo*. The rejection of the traditional method as illegal and the failure to provide any effective alternative have made contemporary international law a bulwark of the existing order to an extent unknown in previous international law or in the municipal law of any civilised country. This is the most fundamental cause of the recent decline of respect for international law ; and those who, in deploring the phenomenon, fail to recognise its origin, not unnaturally expose themselves to the charge of hypocrisy or of obtuseness.

Of all the considerations which render unlikely the general observance of the legal rule of the sanctity of treaties, and which provide a plausible moral justification for the repudiation of treaties, this last is by far the most important. Respect for international law and for the sanctity of treaties will not be increased by the sermons of those who, having most to gain from the maintenance of the existing order, insist most firmly on the morally binding character of the law. Respect for law and treaties will be maintained only in so far as the law recognises effective political machinery through which it can itself be modified and superseded. There must be a clear recognition of that play of political forces which is antecedent to all law. Only when these forces are in stable equilibrium can the law perform its social function without becoming a tool in the hands of the defenders of the *status quo*. The achievement of this equilibrium is not a legal, but a political task

CHAPTER 12

THE JUDICIAL SETTLEMENT OF INTERNATIONAL DISPUTES

BESIDES upholding legal rights, the law provides machinery for settling disputes about these rights. The jurisdiction of national courts is compulsory. Any person cited before a court must enter an appearance or lose his case by default ; and the decision of the court is binding on all concerned.

International law, though it provides machinery for the settlement of disputes, recognises no compulsory jurisdiction. Down to the end of the nineteenth century, the judicial process as applied to international disputes almost invariably took the form of an *ad hoc* agreement to submit a particular dispute to an arbitrator or arbitrators, whose method of appointment was fixed by the agreement and whose verdict was accepted in advance as binding. Under the Hague Convention of 1899, a Permanent Court of Arbitration was established at The Hague. This was, however, not a court, but a standing panel from which suitable arbitrators could be selected by states desiring to resort to arbitration. The Permanent Court of International Justice established under the Covenant of the League of Nations really was a court sitting as such. But it exercised jurisdiction only with the consent of the parties, whether that consent were expressed in an *ad hoc* agreement relating to the particular dispute or in a general agreement between the parties to submit to the Court all disputes falling within certain categories. " It is well established in international law ", declared the Court itself in one of its judgments, " that no state can, without its consent, be compelled to submit its disputes with other states either to mediation or to arbitration or to any other kind of pacific settlement." [1]

Justiciable and Non-justiciable Disputes

In municipal law, all disputes are theoretically justiciable ; for if the point at issue is covered by no legal rule, the answer

[1] *Permanent Court of International Justice*, Series 2, No. 5, p. 27.

of the court will be that the complainant has no case. It is
true that the complainant may not be satisfied with this answer,
and may seek to obtain redress by political action. But this
merely means that he does not want a legal answer, not that
the law has no answer to give, or that the answer is not legally
binding. In international law, all disputes are not justiciable ;
for no court is competent unless the parties to the dispute have
agreed to confer jurisdiction on it and to recognise its decision
as binding. Many treaties are in existence in which the parties
define the kinds of disputes which they agree to recognise as
justiciable as between themselves. In some treaties before 1914,
disputes of certain limited and specific categories were recognised as justiciable. In others, the definition of justiciable
disputes took a negative and somewhat elastic form : the parties
to the treaty undertook to submit to arbitration any dispute
between them which did not affect their " vital interests ",
" independence " or " national honour ". The nearest approach
to a definition of justiciable disputes was contained in Article 13
of the League Covenant, and repeated in Article 36 of the
Permanent Court, which enumerated various kinds of dispute
" declared to be among those which are generally suitable for
submission to arbitration or judicial settlement ". Finally,
several arbitration treaties concluded after 1919, notably those
negotiated at Locarno, recognised as justiciable what were
called disputes between the parties " as to their respective
rights ".

The formulae of the Covenant and the Statute and of the
Locarno arbitration treaties have given a strong impetus to
the idea that international disputes could be classified by an
objective test as *ipso facto* justiciable and *ipso facto* non-
justiciable. Any such classification rests on an illusion. The
formulae in question provide no objective definition of a
justiciable dispute. They merely indicate certain kinds of dispute which the parties to these instruments agree to recognise
as justiciable between themselves. The formula of the Covenant
and the Statute is not really a definition at all, but an enumera-
tion of examples which does not purport to be either exhaustive
or (as the qualification " generally " shews) authoritative.[1]

[1] Disputes " as to the interpretation of a treaty " are the first category of dispute
recognised by the Covenant as " generally suitable " for judicial settlement. It
is noteworthy that the framers of the Covenant, who drew up this article, neverthe-

The Locarno formula is an attempt to give an objective character to the distinction between justiciable and non-justiciable disputes by identifying it with the distinction between conflicts of legal right and conflicts of interest. This formula has little practical value. It merely binds the parties to recognise as justiciable any dispute which they agree to regard as an issue of law. Either party can withdraw any dispute from arbitration by the simple process of placing itself on some other ground than that of legal right. Thus, the British Government, if it had been bound by such a treaty, would presumably have refused to submit to arbitration its default on the war debt agreement with the United States on the ground that the point at issue was not the legal right of the United States to demand payment, and that the dispute was not therefore one as to " respective rights ". As Professor Lauterpacht has conclusively shewn, there is no objective criterion of the " suitability " of a dispute for judicial settlement. " It is not the nature of an individual dispute which makes it unfit for judicial settlement but the unwillingness of a state to have it settled by the application of law." [1] The question which confronts us is twofold : Why are states willing to submit only certain kinds of dispute to judicial settlement, and why do they find it so difficult to define in clear terms what kinds of dispute they are willing to submit ?

The answer to this question must be sought in the necessary relation of law to politics. The judicial settlement of disputes presupposes the existence of law and the recognition that it is binding ; and the agreement which makes the law and which treats it as binding is a political fact. The applicability of judicial procedure depends therefore on explicit or implicit political agreement. In international relations, political agreement tends to be restricted to those spheres which do not affect the security and existence of the state ; and it is primarily

less rejected a proposal to insert in the Covenant a provision that disputes as to its own interpretation should be submitted to the Permanent Court (Miller, *The Drafting of the Covenant*, ii. pp. 349, 516). Behaviour in concrete cases is sometimes more significant than the enunciation *in vacuo* of abstract rules.

[1] Lauterpacht, *The Function of Law in the International Community*, p. 369 and *passim*. It is a pity that Professor Lauterpacht, having brilliantly conducted his analysis up to the point where the unwillingness of states is recognised as the limiting factor in the justiciability of international disputes, should have been content to leave it there, treating this " unwillingness ", in true utopian fashion, as perverse and undeserving of the attention of an international lawyer.

in these spheres that the judicial settlement of disputes is
effective The majority of international disputes which have
in the past been settled by arbitration or by some other legal
procedure have been either pecuniary claims or disputes about
national frontiers in remote and sparsely inhabited regions.
The exclusion, in arbitration treaties concluded before 1914,
of disputes affecting " vital interests ", " independence " or
" national honour " meant the exclusion of precisely those
matters on which political agreement could not be attained.
When political disagreement threatened, arbitration was recognised as impracticable. We shall see presently that what is
virtually the same reservation was maintained in subsequent
agreements for arbitration or judicial settlement in the form
of the exclusion from these agreements of disputes endangering
the sanctity of existing treaties or existing legal rights.

The same consideration explains why no definition of disputes recognised as justiciable can be universally or permanently
valid ; for political agreement is a factor which varies from
place to place and from time to time.[1] Prior to 1917 there was
a general political understanding throughout the world that
the property rights of individuals were valid, and that a foreigner
whose property was for any reason confiscated by the government of the country in which it was situated had a claim in
international law to compensation. So long as this understanding existed, claims based upon it could be settled by
arbitration. With the establishment of the Soviet régime in
Russia, this understanding ceased to apply to that country ;
and when the Soviet Government made its first important
international appearance at the Genoa Conference in 1922, it
was careful to scout in advance the idea that property claims
against it should be submitted to arbitration. " In the trial
of disputes of this kind ", ran the memorandum which it submitted to the Conference, " the specific disagreements will
inevitably end in opposing to one another two forms of property.
. . . In such circumstances there can be no question of an
impartial super-arbiter." And when, at the subsequent Hague
Conference, the British delegate pathetically enquired " whether

[1] The British Government, in its memorandum of 1928 on arbitration (*League
of Nations : Official Journal*, pp. 694-704), criticised general arbitration treaties
on the ground that, in the case of every country, " obligations which it may be
willing to accept towards one state it may not be willing to accept towards another ".

it would be impossible to find a single impartial judge in the whole world ", Mr. Litvinov firmly replied that " it was necessary to face the fact that there was not one world, but two, a Soviet world and a non-Soviet world ".[1] " Impartiality " is a meaningless concept where there is no common ground at all between the two contending views. Judicial procedure cannot operate without accepted political postulates.

The assumption of the British delegate just quoted that the obstacle to international arbitration was the difficulty of finding impartial judges had been heard on previous occasions. " The great obstacle to the extension of arbitration ", declared the American delegate at the Hague Conference of 1907, " is not the unwillingness of civilised nations to submit their disputes to the decision of an arbitral tribunal ; it is rather an apprehension that the tribunal selected will not be impartial." Lord Salisbury is quoted in a similar sense.[2] This opinion rests on a misapprehension. The potential personal bias of the international judge is not the real stumbling-block. The popular prejudice against submitting matters of national concern to the verdict of a " foreigner " is based primarily, not on the belief that the foreign judge will be biassed as between the parties, but on the fact that there are certain fundamentals of a political character which we are not prepared to have challenged by any foreign authority, whether judicial or political. The abolition of private ownership for Soviet Russia, the right of blockade for Great Britain, the Monroe Doctrine for the United States are familiar examples of such political fundamentals. Such fundamentals need not, however, be major issues at all. Palmerston treated the Don Pacifico episode in 1850, and Signor Mussolini the murder of an Italian general in Greece in 1923, as political issues which they were not prepared to submit to judicial settlement.[3]

But there is another and more general sense in which the

---

[1] Quoted in Taracouzio, *The Soviet Union and International Law*, p. 296.

[2] *Proceedings of the Hague Peace Conference* (Engl. transl. : Carnegie Endowment), *Conference of 1907*, ii. p. 316.

[3] On the latter occasion, Professor Gilbert Murray, representing South Africa on the Assembly of the League of Nations, lamented that a judicial question (i.e. compensation for Italy) had been brought before a political organ and decided on political grounds (*League of Nations : Fourth Assembly*, pp. 139 *sqq.*) — an excellent example of the fallacy, so trenchantly exposed by Professor Lauterpacht, that certain issues are *ipso facto* judicial.

absence of common political presuppositions impedes the development of the judicial process in the international community. Municipal law, though far more fully and minutely developed than international law, is never wholly self-sufficing. The application of the law to the particular case is always liable to involve an element of judicial discretion, since the legislator can hardly have foreseen all the relevant circumstances of every case arising under the law. " There are many situations ", writes Dean Pound, " where the course of judicial action is left to be determined wholly by the judge's individual sense of what is right." [1] It would perhaps have been fairer to say that the good judge will be guided in such cases not so much by his own " individual sense of what is right " as by the sense of right generally accepted by the community whose servant he is. But that some "sense of what is right", whether individual or general, is a necessary ingredient of many judicial decisions, few will care to deny. The importance of the political presuppositions which inspire the Supreme Court of the United States in the interpretation of the Constitution, and the way in which, in the course of American history, these presuppositions have changed in response to changing social conditions, is well known.[2] The problem is, in its final analysis, the fundamental one of the relation of the rights of the individual to the needs of the community. Every national community has necessarily found a working solution of this problem. The international community has not yet done so. The controversy about the freedom of the seas shews that Great Britain would be unwilling to risk any interpretation of her maritime rights by an international court in the light of the supposed needs of the international community as a whole ; and there are important matters on which every other Great Power would make similar reservations. The absence of an accepted view of the general good of the community as a whole overriding the particular good of any individual member of it, which we have already noted as the crucial problem of international morality,

[1] Roscoe Pound, *Law and Morals* (2nd ed.), p. 62.
[2] Professor Laski remarked many years ago that " the foreigner in the United States cannot but observe with the deepest wonder how eagerly possible nominations for a vacant position on the Supreme Court are canvassed " (Introduction to English translation of Duguit, *Law in the Modern State*, xxiii.). The wonder has grown less deep since the political character of the Court has been better understood.

also stands in the way of the development of judicial settlement in its application to international disputes.

We find, therefore, in the problem of the justiciability of international disputes another illustration of the fact that law is a function of political society, is dependent for its development on the development of that society, and is conditioned by the political presuppositions which that society shares in common. It follows that the strengthening of international law, and the extension of the number and character of international disputes recognised as suitable for judicial settlement, is a political, not a legal, problem. There is no principle of law which enables one to decide that a given issue is suitable for treatment by legal methods. The decision is political ; and its character is likely to be determined by the political development of the international community or of the political relations between the countries concerned. Similarly, there is no principle of law which enables one to decide whether a rule of law or a legal institution which has proved its value in a national community should be introduced by analogy into international law. The sole valid criterion is whether the present stage of political development of the international community is such as to justify the introduction of the rule or institution in question. In modern international relations, the machinery of judicial settlement has been developed far in advance of the political order in which alone it can effectively operate. Further progress towards the extension of the judicial settlement of international disputes can be made, not by perfecting an already too perfect machinery, but by developing political co-operation. The fact that the members of the British Commonwealth of Nations have hitherto steadfastly refused to set up any kind of permanent and obligatory procedure for the judicial settlement of disputes between one another should serve as a warning to those who are disposed to attach undue importance to the perfection of judicial machinery in international relations. It is a curious paradox that, by signing the Optional Clause of the Statute of the Permanent Court and by excluding from its operation inter-Commonwealth disputes, Great Britain and Dominions are bound in this respect towards many foreign countries by an obligation more far-reaching than they have assumed among themselves.

Projects of "All-in Arbitration"

Many thinkers of the period between the two wars went, however, far beyond mere plans for the modest and gradual extension of the scope of judicial procedure in international relations. It became a widely cherished ambition to provide, by a stroke of the pen, for the compulsory settlement of all international disputes by arbitration. Schemes for obligatory arbitration were mooted on many occasions prior to 1914, but failed to win acceptance. The Covenant of the League of Nations, while providing for the establishment of the Permanent Court and encouraging the submission of suitable disputes to arbitration or judicial settlement, gave little encouragement to the advocates of obligatory arbitration. In all disputes, it left the choice of the procedure to the discretion of the states concerned ; and the political procedure of " enquiry by the Council " always remained open. It was precisely this political aspect of the Covenant which became a target for the attacks of the utopian school. A widespread feeling grew up that the way to establish an international " rule of law " and avoid future wars was for states to submit all international disputes of every kind to an international arbitral tribunal having power to decide them at its discretion on grounds either of strict law or of equity and common sense. Such was the vague conception summed up in the popular catchword of " all-in arbitration ". This demand for " all-in arbitration " was supposed to have been met by the Geneva Protocol and by the General Act. It was widely believed that, had the British Government not rejected the Protocol, or had the General Act been accepted without reservations by the principal Powers, a satisfactory procedure would have been in existence for the compulsory arbitration of all international disputes and an important cause of war removed.

But here we come upon an extraordinary confusion, or series of confusions, of thought which, throughout this period, enveloped and obscured the problem of the peaceful settlement of international disputes When the League Covenant, by an amendment inserted after the establishment of the Permanent Court of International Justice, set " judicial settlement " side by side with " arbitration ", " arbitration " meant the verdict of a judge or a tribunal appointed *ad hoc*, and " judicial settlement " the verdict of a regularly constituted court ; and there is no reason to suppose that any other distinction was intended between them. But the misguided attempt to discover an objective distinction between justiciable and non-justiciable disputes led to an equally fallacious distinction between " judicial settlement ", meaning the settlement of " justiciable " disputes in accordance with the letter of the law, and " arbitration ", meaning the settlement of " non-justiciable " disputes, which were not covered by the letter of the law, on grounds of equity. This conception left its traces on the Geneva Protocol. According to the Assembly report on that instrument, " the arbitrators need not necessarily be jurists ", and if they obtain an advisory opinion on any point of law from the Permanent Court, that opinion is " not legally binding on them ".[1] But the distinction between " judicial settlement " and " arbitration " was first fully developed in the General Act. Under this instrument, disputes " with regard to which the parties are in conflict as to their respective rights " were to be referred to the Permanent Court for " judicial settlement ". All other international disputes were to be referred for "arbitration " to an arbitral tribunal. In the absence of any agreed stipulation to the contrary, the tribunal, in pronouncing its judgment was to apply the same rules of law as were applied by the Permanent Court. But " in so far as there exists no such rule applicable to the dispute, the tribunal shall decide *ex aequo et bono* ". This reference to rules of law seems incomprehensible. If the dispute turned on legal rights, it would be submitted not to the arbitral tribunal, but to the Permanent Court. If it did not turn on legal rights, the dispute could not be solved by the application of legal rules. The conception that there is a class of international disputes which arise, so to speak, *in vacuo*, and are not affected by any existing legal rights or by any rule of international law, is a pure myth.

A more serious confusion is, however, in store. There is a perfectly valid distinction, familiar both in national and in international affairs, between " legal " disputes, arising out of claims which purport to be based on existing legal rights, and " political " disputes arising out of claims to alter existing legal rights. The difference turns, however, not on the nature of the dispute, but on the question whether the complainant seeks

[1] *League of Nations : Fifth Assembly*, First Committee, p. 486.

his remedy through legal or through political procedure. In the state, claims of the former kind are dealt with by the courts, claims of the latter kind by political action. The individual who fails to get his grievance remedied by a court may seek a remedy for the same grievance through legislation. Internationally, the distinction is less clear cut. No international court is recognised as competent to settle all " legal " disputes, and there is no recognised machinery to settle all " political " disputes. In these circumstances, states making claims against other states are not obliged to make it clear, and do not always make it clear, whether the claim is based on legal rights or is tantamount to a demand to alter those rights. But the distinction, though sometimes obscured in practice, is real enough. Both nationally and internationally, " political " disputes are, generally speaking, more serious and more dangerous than " legal " disputes. Revolutions and wars are less likely to arise from disputes about existing legal rights than from the desire to change those rights. The wise politician, and the wise student of politics, will devote a great deal of attention to political disputes.

When, therefore, it was officially claimed that the Geneva Protocol constituted " a system for the pacific settlement of *all disputes* which might ever arise ",[1] or that the General Act provided " a comprehensive method of settling all international disputes of whatever character ",[2] the conclusion might reasonably have been drawn, and was in fact drawn by many people, that provision had been made for the settlement by arbitration of political disputes, i.e. of disputes arising from claims to modify existing legal rights. Closer inspection did not, however, justify this conclusion. In an inconspicuous passage of the Assembly report on the Protocol, it was explained that the procedure did not apply to " disputes which aim at revising treaties and international acts in force or which seek to jeopardise the existing territorial integrity of signatory states ". In fact, added the *rapporteur*, " the impossibility of applying compulsory arbitration to such cases was so obvious that it was quite superfluous to make them the subject of a special provision ".[3] The General Act is less ingenuous. It

[1] *League of Nations : Fifth Assembly*, p. 497.
[2] *Memorandum on the General Act*, Cmd. 3803, p. 4.
[3] *League of Nations : Fifth Assembly*, p. 194.

purports to enforce compulsory arbitration for disputes which are not disputes about the " respective rights " of the parties. It purports to authorise the arbitral tribunal to decide such disputes *ex aequo et bono.* But the authorisation applies only " in so far as there exists no [legal] rule applicable to the dispute " ; and this qualification has the same effect as the reservation in the report on the Geneva Protocol. The essence of a political dispute is the demand that the relevant legal rule, though admittedly applicable, shall not be applied. When a dispute arises through the claim of a state that its existing frontiers, or existing treaty restrictions on its sovereignty, or existing obligations under a financial agreement, are intolerable, it is useless to refer it to an arbitral tribunal whose first duty is to apply the legal " rule applicable to the dispute ". The legal right exists and is uncontested. The dispute arises from a demand to change it. Political disputes cannot be settled within the framework of the law by tribunals applying rules of law. The Geneva Protocol and General Act, though purporting to provide for the peaceful settlement of all international disputes, in fact left the most important and dangerous category of international disputes untouched.

No scheme of " all-in arbitration " more inclusive than the make-believe of the Geneva Protocol and the General Act was officially propounded or considered. Some governments were prepared to accept arbitration for such disputes as did not endanger the existing political order — a limitation hardly less restrictive than the vital interests, independence and national honour of the older arbitration treaties. But no government was willing to entrust to an international court the power to modify its legal rights. Some theorists, however, were more ready than practical statesmen to brush this difficulty aside, and were quite prepared to entrust to a so-called arbitral tribunal the task not only of applying existing rights, but of creating new ones. A British organisation called the New Commonwealth Society evolved an elaborate scheme for an arbitral tribunal which would " determine, on the basis of equity and good conscience, political disputes, including those which have to do with the revision of treaties ", thus establishing " an indirect method of legislation in the affairs of nations " by an equity tribunal.[1] Such a scheme would appear to be

---

[1] Lord Davies, *Force*, pp. 73, 81.

the necessary corollary of Professor Lauterpacht's belief that international " conflicts of interests are due . . . to the imperfections of international legal organisation ".[1] International conflicts of interests will in future be resolved by a tribunal which will become the supreme organ of world government, exercising not merely the judicial function of interpreting the rights of states, but the legislative function of changing them. Thus will be realised another distinguished international lawyer's dream of " an international legal community whose centre of gravity is in the administration of international justice ".[2]

These theories have one important merit. They recognise the fallacy, implicit in the Geneva Protocol and the General Act, that an international legal order based on the recognition, interpretation and enforcement of existing rights is an adequate provision for the peaceful settlement of international disputes. But in avoiding this fallacy, they fall into a still graver one. Perceiving that provision must be made for the modification of existing rights, they force this essentially political function into a legal mould and entrust its exercise to a tribunal. Unwilling to recognise the political basis of every legal system, they dissolve politics into law. In this quasi-judicial twilight, the judge becomes the legislator, political issues are settled by an impartial tribunal on grounds of equity and common sense, and the distinction between law and politics disappears.

The extreme difficulty of the international problem is no doubt responsible for the prescription of so heroic a remedy. But the fact that the problem is difficult scarcely justifies us in propounding a solution which nobody regards as either feasible or desirable in our far more highly organised national communities. The obligatory arbitration of international disputes of all kinds is, according to Professor Lauterpacht, " a *sine qua non* of the normal machinery for the preservation of peace ".[3] Yet obligatory arbitration of claims not based on legal right is rarely enforced in civilised states, and least of all in those which enjoy the longest record of domestic peace. It does not occur to us to attribute " conflicts of interests " in our domestic politics to the imperfections of our legal organisation, or to

[1] Lauterpacht, *The Function of Law in the International Community*, p. 250.
[2] Kelsen, *The Legal Process and International Order*, p. 30.
[3] Lauterpacht, *The Function of Law in the International Community*, p. 438.

submit to a national arbitral tribunal, for impartial decision on grounds of equity and common sense, disputes about the necessity of conscription, the abolition of the means test, the legal status of trade unions, or the nationalisation of mines. The difficulty is not that we could not find a group of impartial persons deeply imbued with the principles of equity and common sense, but that impartiality, equity and common sense are not the primary, or at any rate not the sole, qualities which we require in a decision of such issues. These are political issues, and are settled by procedure which allows for the intrusion of power, whether in the form of a majority vote, as in democracies, or of the will of a dictator or a party, as in authoritarian states. Neither in democracies nor in authoritarian states are such issues decided by an " impartial " tribunal.

The Inapplicability of Judicial Procedure to " Political " Disputes

Why then is it necessary, not only in theory, for the sake of clear thinking, but also in practice, for the sake of good government, to preserve this distinction between the legal and political, between issues which we are willing to have settled by judicial procedure on grounds of existing legal rights, and issues which can only be settled by political procedure because they turn on a demand for the modification of existing legal rights ?

The first answer is that judicial procedure differs fundamentally from political procedure in excluding the factor of power. When a dispute is submitted to a court, the presupposition is that any difference in power between the parties is irrelevant. The law recognises no inequality other than inequality of legal right. In politics, the converse presupposition holds. Here power is an essential factor in every dispute. The settlement of a conflict of interest between British agriculturalists and British industrialists will depend, in part at any rate, on their respective voting strength and the respective " pulls " which they can exercise on the government. The settlement of a conflict of interest between the United States and Nicaragua will depend, in greater part (for the ratio of power to other factors is higher in international than in national politics), on the relative strength of the two countries. Conflicts of interest can be dealt with only by an organ which takes the power factor into account. Nothing is gained, and the proper function of law is debased and discredited, if this political function is entrusted to a tribunal whose constitution and procedure are deliberately assimilated to those of a court of law. As Mr. Bernard Shaw has remarked, the functions of judge and legislator are " mutually exclusive " : the former must ignore every interest, the latter take every interest into account.[1]

The second answer is equally fundamental. We have seen that even the strictly judicial procedure of a court sometimes entails political presuppositions, if only because the application of the law to the particular case is always liable to involve an element of judicial discretion, and this discretion, if it is not to be purely capricious, must draw its inspiration from those presuppositions. Where a tribunal is called upon to decide not on issues of legal right, but on claims to set aside legal rights in favour of equity or common sense, the necessity of clearly defined political pre-suppositions becomes all the more obvious. In such cases, judicial discretion, instead of being limited to points left ambiguous by the law, has infinite scope ; and the decisions of the tribunal, if they are not to be mere expressions of individual opinion, must be based on well-established assump-tions shared by the community as a whole or by those who speak in its name. The existence of such assumptions in national communities sometimes makes possible the use of arbitration even in political issues ; and the same possibility is not entirely excluded in the international sphere. But generally speaking, it is a fundamental obstacle to international arbitration *ex aequo et bono* that common assumptions of a far-reaching kind scarcely exist in the international community. To submit to an inter-national tribunal, for decision on grounds not of law, but of equity and common sense, disputes concerning British interests in Egypt or the interests of the United States in the canal zone of Panama, or the future of Danzig, or the frontiers of Bulgaria, would have been impracticable, not only because the settlement of these problems involves issues of power, but also because there is no political agreement even of the vaguest kind as to what equity and common sense mean in relation to such questions. On the rare occasions on which international

[1] G. B. Shaw, *John Bull's Other Island*, Preface.

tribunals have been empowered by the parties to decide issues between them on grounds other than those of strict law, the tribunals have shewn the greatest reluctance to avail themselves of the discretion accorded to them ; not, as Professor Lauterpacht supposes, because " law is more just than loose conceptions of justice and equity ",[1] but because no responsible tribunal cares to commit itself on any important issue to an authoritative pronouncement as to what is " equitable " or " just " in international relations. An international tribunal, once it has left the comparatively solid ground of international law and legal rights, can find no foothold in any agreed conception of equity or common sense or the good of the community. It remains, in Professor Zimmern's words, " an array of wigs and gowns vociferating in emptiness ".[2]

The crux, however, remains. Political issues, both nationally and internationally, are far more menacing than issues of legal right. The periodical, or rather the constant, revision of existing rights is one of the prime necessities of organised society ; and to bring about revision in the international society by means other than war is the most vital problem of contemporary international politics. The first step has been to extricate ourselves from the blind alley of arbitration and judicial procedure, where no solution of this problem is to be found. Having taken this step, we are free to approach it by other, and perhaps more promising, avenues.

[1] Lauterpacht, *The Function of Law in the International Community*, p. 252.
[2] Zimmern, *The League of Nations and the Rule of Law*, p. 125. The words are applied to Taft's international arbitral court. They could be applied, still more appositely, to the equity tribunal advocated by the New Commonwealth Society.

CHAPTER 13

PEACEFUL CHANGE

RECOGNITION of the need for political change has been a commonplace of thinkers of every period and every shade of opinion. " A state without the means of some change ", said Burke in a famous phrase, " is without the means of its own conservation." [1] In 1853, Marx wrote trenchantly on the Eastern question :

> Impotence . . . expresses itself in a single proposition : the maintenance of the *status quo*. This general conviction that a state of things resulting from hazard and circumstances must be obstinately maintained is a proof of bankruptcy, a confession by the leading Powers of their complete incapacity to further the cause of progress and civilisation.[2]

And Professor Gilbert Murray has put the same point in a different form :

> War does not always arise from mere wickedness or folly. It sometimes arises from mere growth and movement. Humanity will not stand still.[3]

It appears to follow from this view that the attempt to make a moral distinction between wars of " aggression " and wars of " defence " is misguided. If a change is necessary and desirable, the use or threatened use of force to maintain the *status quo* may be morally more culpable than the use or threatened use of force to alter it. Few people now believe that the action of the American colonists who attacked the *status quo* by force in 1776, or of the Irish who attacked the *status quo* by force between 1916 and 1920, was necessarily less moral than that of the British who defended it by force. The moral criterion must be not the " aggressive " or " defensive " character of the war, but the nature of the change which is being sought and resisted. " Without rebellion, mankind

---

[1] Burke, *Reflexions on the Revolution in France* (Everyman ed.), p. 19.
[2] Marx and Engels, *Works* (Russian ed.), ix. p. 372.
[3] Gilbert Murray, *The League of Nations and the Democratic Idea*, p. 16.

would stagnate and injustice would be irremediable." [1] Few serious thinkers maintain that it is always and unconditionally wrong to start a revolution ; and it is equally difficult to believe that it is always and unconditionally wrong to start a war. Everyone will, however, agree that war and revolution are undesirable in themselves. The problem of " peaceful change " is, in national politics, how to effect necessary and desirable changes without revolution and, in international politics, how to effect such changes without war.

Every effective demand for change, like every other effective political force, is compounded of power and morality ; and the object of peaceful change can be expressed in terms neither of pure power nor of pure morality. It is rather unprofitable, except as an academic exercise, to enquire whether the purpose of any change should be to establish " justice ", by remedying " just " grievances, or to maintain " peace ", by giving satis-faction to those forces which would otherwise be strong enough to make revolution or war. But it is dangerous to suppose that the two purposes are identical, and that no sacrifice of one or the other is required. Every solution of the problem of political change, whether national or international, must be based on a compromise between morality and power.

The Role of Power in Political Change

The necessary role of power in political change will be ignored only by the most superficial observers. Few " revision-ist " campaigns in history have been more firmly based on moral considerations than that of the Dreyfusards in France. Yet the protest against the condemnation of Dreyfus would never have been effective if it had not been taken up by powerful political organisations and used by them as a weapon against political opponents. The grievances of Albania and Nicaragua, whatever their moral basis, will never be effective unless they are endorsed, for interested reasons, by some Great Power or Powers. It is fair to attribute the growth of social legislation in the last hundred years to a growing realisation of the just grievances of the working class. Yet these results would never have been achieved without the constant use, or threatened use, of force in the form of strikes and revolutions. " It is true ",

[1] B. Russell, *Power*, p. 263.

remarks Mr. John Strachey, " that governments always tell
us that they will never yield to force. All history tells us, however, that they never yield to anything else." [1] " Peaceful
secession ! " exclaimed Daniel Webster in 1849. " Sir, your
eyes and mine are never destined to see that miracle." [2] " The
winning back of the lost territories ", wrote Hitler in a famous
passage of *Mein Kampf*, " is not achieved through solemn
invocations of the Lord God or through pious hopes in a League
of Nations, but through armed force." [3] Hitler might even
have appealed to the respectable authority of Gladstone who,
in the days when liberalism was still a political force, observed
that " if no considerations in a political crisis had been addressed
to the people of this country except to remember to hate violence
and love order and exercise patience, the liberties of this country
would never have been obtained ".[4] It has been said that
no ruling class ever abdicates of its own accord. Article 19 of
the Covenant of the League of Nations remains a lonely monu-
ment to the pathetic fallacy that international grievances will
be recognised as just and voluntarily remedied on the strength
of " advice " unanimously tendered by a body representative
of world public opinion.

While, however, the fundamental problem of political change
— the compromise between power and morality — is identical
in national and in international politics, the question of procedure is complicated by the unorganised character of the
international community. Analogies drawn from procedures
of change in the national sphere can only be applied with
caution to the international field. We have seen that judicial
procedures cannot be invoked, either nationally or internationally, for the solution of ultimate political problems. But the
analogy of legislation seems at first sight more hopeful. The
legislative process, unlike the judicial process, recognises the
role of power which is inherent in all political change (for the
legislative authority is the supreme power of the state imposing
its will on the whole community) ; and legislation, which a
German writer has called " legal revolution ",[5] is the most
obvious and regular way of bringing about political change

[1] J. Strachey, *The Menace of Fascism*, p. 228.
[2] Quoted in J. Truslow Adams, *The Epic of America*, p. 239.
[3] Hitler, *Mein Kampf*, p. 708.
[4] Quoted in E. Pethick-Lawrence, *My Part in a Changing World*, p. 269.
[5] Berber, *Sicherheit und Gerechtigkeit*, p. 9.

Peaceful Change

within the state. " What is peaceful change as an effective
institution of international law or of international society ? "
asks Professor Lauterpacht, and answers : " It is the acceptance
by states of a legal duty to acquiesce in changes in the law
decreed by a competent international organ ".[1]

It has already been noted that international law rests upon
custom, and that there is at the present time no such thing as
international legislation or an international legislature. The
terms of Article 19 of the Covenant shew how remote the
principal states were in 1919 from " acceptance of a legal
duty to acquiesce in changes in the law decreed by a competent international organ ". Nor can this well be otherwise.
Reflexion will shew that the legislative process, like the judicial
process, presupposes the existence of a political order. It is
only by that combination of consent and coercion which under-
lies every political society that we can arrive at the establishment
of a supreme organ, whether parliament or council of state
or individual autocrat, whose fiat creates law binding on all
members of the community. These conditions are not fulfilled
in the international community. The Assembly of the League
of Nations, whose decisions required unanimity, was a conference empowered to conclude international agreements, not
a legislature which passed international legislation ; for, as
Mr. Eden bluntly observed at one of its sessions, " it would
plainly be impracticable . . . to give the Assembly power to
impose changes against the wish of the parties concerned ".[2]
The difficulty lies not in the lack of machinery for international
legislation, but in the absence of an international political order
sufficiently well integrated to make possible the establishment
of a legislative authority whose decrees will be recognised as
binding on states without their specific assent. If we accept
Professor Lauterpacht's identification of peaceful change with
international legislation, we can only conclude that, in his
words, " an international system of peaceful change . . . runs
the risk of being unreal unless it forms part and parcel of a
comprehensive political organisation of mankind ".[3] The condition of international legislation is the world super-state.

Need we, however, reconcile ourselves to the discouraging

---

[1] *Peaceful Change*, ed. C. A. W. Manning, p. 141.
[2] *League of Nations : Seventeenth Assembly*, p. 46.
[3] *Peaceful Change*, ed. C. A. W. Manning, p. 164.

conclusion that any international system of peaceful change must await the coming of the super-state ? The analogy of legislation may turn out to be not merely discouraging, but misleading. The present almost universal belief in the beneficence of legislation as a reforming instrument within the state is in the main a growth of the past fifty years. Down to the end of the nineteenth century, many intelligent people continued to regard the state as a necessary evil and legislation as a regrettable device not to be resorted to except in case of proven necessity.[1] Within the national community, the distinction familiar to nineteenth-century thought between " society " and " state " has lost much of its significance through the development of the social functions of the modern state. But in the international sphere, we are in the presence of a " society " which has no corresponding " state " ; and we may therefore find some help in the conception, which would hardly have seemed paradoxical to any age but our own, of changes peacefully effected in the social structure without legislation or any other overt form of state intervention. Even to-day, it is easy to exaggerate the role of legislation ; and it may still be true to say (as it would certainly have been true a hundred years ago) that the most important changes in the structure of society and in the balance of forces within it are effected without legislative action. It may be unnecessarily pessimistic to rush into the conclusion that the absence of an international legislature rules out any international procedure of peaceful change.

If, therefore, we are looking for the nearest analogy in the national community to the turbulent relations which render the problem of change acute in the international society, we may find it in the relations of those group-entities within the state whose conflicts have not been in the past, and still in large measure are not, settled by any legislative process. Of these group-entities, by far the most important, and by far the most instructive for our purpose, are those representing capital and labour respectively. Here we have the same recurrent conflict between " haves " and " have-nots ", between " satisfied " and " dissatisfied " ; the same reluctance, on the part of one or

---

[1] Professor Lauterpacht's remark that " the circle of interests directly regulated by law expands with the growth of civilisation " (*The Function of Law in the International Community*, p. 392) is a truism to-day, but would have seemed a paradox to many nineteenth-century thinkers.

both sides, to accept the principle of " all-in arbitration " for the settlement of their disputes ; the same recognition of the inapplicability or inadequacy of the legislative process ; the same appeals to " law and order " by the satisfied group; and the same use, or threatened use, of violence by the dissatisfied in order to assert their claims. It is sometimes said that there can be no international procedure of peaceful change so long as states insist on being judges in their own cause. Here is a class of disputes in which both parties commonly insist on being judges in their own cause, and in which some progress at least has been made towards an orderly procedure of peaceful change.

Force has always been a crucial factor in relations between capital and labour. In the beginnings of the industrial revolution, every attempt at organised self-help on the part of the workers was rigorously repressed. This unqualified repression ended in Great Britain with the repeal of the Combination Acts in 1825, and continued in Russia as late as 1905. Between those two dates, the workers of every important industrial country secured recognition of their right to use the weapon of the organised strike. The strike not only proved itself an effective instrument for extracting concessions from employer to workers, but became a recognised symbol of the major weapon of force — revolution.[1] In recent times, the element of force has been once more eliminated from relations between employer and workers by the authoritarian governments of Soviet Russia,[2] Italy and Germany, through legislation prohibiting strikes and an executive strong and ruthless enough to enforce the prohibition. Democratic countries have from time to time prohibited strikes, though such prohibitions have nearly always been resisted by the workers, and have rarely proved enforceable over an extended period.[3] Theoretically,

[1] This is the significance of the " one-day strike ", which was popular in certain countries and which, though useless in itself, was designed to demonstrate that the workers were strong enough to break the power of the state. The success or failure of the one-day strike was thus a test of power, and its result enabled both sides to draw the appropriate conclusion without resorting to extreme measures.

[2] The position is not substantially affected by the fact that in Soviet Russia the employer is normally a state trust or institution.

[3] In Great Britain, strikes in munitions factories were prohibited during the first world war by the Munitions of War Acts. But though strikes occurred, the law was rarely if ever enforced, and it came to an end with the War. Under the Trade Disputes Act of 1927, political strikes were declared illegal, but no such case appears to have arisen since the act was passed. The situation in other

force might in the same way be eliminated from the settlement of international disputes by a powerful and authoritarian super-state. But this result, whether desirable or not, lies outside the scope of practical consideration ; and we shall therefore find a better analogy to the international position if we consider those countries and those periods in which relations between capital and labour have not been dominated by the over-whelming power of the state. In the latter part of the nineteenth century and the first part of the twentieth the " have-nots " of most countries steadily improved their position through a series of strikes and negotiations, and the " haves ", whether through a sense of justice, or through fear of revolution in the event of refusal, yielded ground rather than put the issue to the test of force. This process eventually produced on both sides a willingness to submit disputes to various forms of conciliation and arbitration, and ended by creating something like a regular system of " peaceful change ". In many countries such a system has been in operation for many years with remarkable success, though the ultimate right to resort to the weapon of the strike is not abandoned. If we could apply this analogy to international relations, we might hope that, once the dis-satisfied Powers had realised the possibility of remedying grievances by peaceful negotiations (preceded no doubt in the first instance by threats of force), some regular procedure of " peaceful change " might gradually be established and win the confidence of the dissatisfied ; and that, once such a system had been recognised, conciliation would come to be regarded as a matter of course, and the threat of force, while never form-ally abandoned, recede further and further into the background. Whether the analogy is in fact valid, or whether this hope is purely utopian, is a question which can hardly be settled except by the test of experience. But one may record with some confidence the view that this is the only line of advance which affords any prospect at all of the establishment of any inter-national procedure, however imperfect, of peaceful change.

The implication of this procedure must, however, be clearly recognised. Few issues of social or political change of sufficient

countries is summarised in a pamphlet published by the American League for Industrial Democracy (*Shall Strikes be Outlawed ?* by Joel Seidman), which con-cludes that " labour feels that its right to strike is its surest guarantee of fair treat-ment " and that " along the path of voluntary collective bargaining lies the greatest hope of satisfactory industrial relations ".

magnitude to involve the risk of revolution or war can be settled without detriment, or apparent detriment, to the interests of one of the parties. That the party at whose expense the change was to be effected would acquiesce in it without the existence of means of pressure to compel him to do so was one of the strange illusions of the ill-fated Article 19 of the Covenant; and this illusion may be discarded. Such self-abnegation could indeed hardly have been expected. The statesman, the trade union leader or the company director is a trustee for those whose interests he represents; and in order to justify extensive concessions at their expense, he must generally be in a position to plead that he has yielded to *force majeure*. When the change is effected by legislation, the compulsion is that of the state. But where the change is effected by the bargaining procedure, the *force majeure* can only be that of the stronger party. The employer who concedes the strikers' demands pleads inability to resist. The trade union leader who calls off an unsuccessful strike pleads that the union was too weak to continue. "Yielding to threats of force", which is sometimes used as a term of reproach, is therefore a normal part of the process.

The parallel should not be pressed too far. The role of force, even in the most advanced democratic states, is indeed more constant and more conspicuous than most sentimental democrats care to admit. In so orderly a country as Great Britain, during the present century, force has been used or threatened for securing political ends by Ulstermen, Irish nationalists, female suffragists, communists, fascists and organised workers. But within the state there are checks on the too hasty resort to force. In the first place, the legislative process exists, and provides an alternative method of change; faith in the ballot-box has deterred the workers of many countries from revolutionary policies. Secondly, the state makes some show (often an imperfect one) of holding the balance impartially between the parties on the issue in dispute. Thanks to these checks, a certain moral discredit attaches in democratic countries, in the minds of all classes, to the open use or threat of force until other means have been tried of bringing about a change.

In international politics neither of these checks exists. The use or threatened use of force is therefore a normal and recognised method of bringing about important political change, and is regarded as morally discreditable mainly by those " conservative " countries whose interests would suffer from change. The largest operation of " peaceful change " in the nineteenth century was that performed by the Congress of Berlin, which revised the treaty imposed by Russia on Turkey at San Stefano. But this revision took place only under the tacit threat of a declaration of war against Russia by Great Britain and Austria-Hungary.[1] The Lausanne Treaty of 1923 was a revision, extorted by the use and threatened use of force, of the treaty signed with Turkey at Sèvres in 1920. It was denounced by Lloyd George as " an abject, cowardly and infamous surrender " ; and this opinion was widely held at the time.[2] The revision of Czecho-Slovakia's frontiers effected by the Munich Agreement of September 1938 was also the product of a threat of force ; and here we have the explicit testimony of M. Benes that no alternative method was available. For five years earlier he had publicly stated " that no country could be forced by anyone to revise its frontiers, and that anyone who attempted it in the case of Czecho-Slovakia would have to bring an army along ".[3] Another curious example may be added. When Poland annexed Vilna in 1920, Lithuania closed the frontier and severed all communications with Poland. It is doubtful whether this isolation conferred any advantage on Lithuania. But no Lithuanian statesman could have justified the reopening of the frontier and the consequent loss of face to his country if he had not been in a position to plead *force majeure*. In March 1938, Poland mobilised an army and presented an ultimatum to Lithuania. The frontier was at once reopened ; and normal relations were established. An operation of peaceful change, generally recognised as salutary, could not be effected save under a threat of war. Normally, the threat of war, tacit or overt, seems a necessary condition of important political changes in the international sphere.[4]

[1] A writer who has surveyed the history of peaceful change down to 1914 records the conclusion that " it is always wisest to face Europe with a *fait accompli* " (Cruttwell, *History of Peaceful Change*, p. 3).

[2] D. Lloyd George, *The Truth About the Peace Treaties*, ii. p. 1351.

[3] *The Times*, April 26, 1933, quoted by Professor Manning in *Politica*, December 1938, p. 363.

[4] Those who assert that change effected under a threat of armed force is not " peaceful change " are, of course, at liberty to define their terms how they please. But it should be noted that a definition thus restricted would equally exclude changes effected by a legislative or judicial process, if these required enforcement.

This principle has not only been demonstrated in practice on many occasions, but received a large measure of theoretical recognition from the framers and interpreters of the Covenant of the League of Nations. The machinery of the League was brought into action by the danger of war. Article 11 related to " any war or threat of war " and to " any circumstance . . . which threatens to disturb international peace "; and Article 19 purported to deal with "treaties which have become inapplicable " (a phrase which has never been satisfactorily explained) and with " international conditions whose continuance might endanger the peace of the world ". Moreover, the most effective article of the Covenant for promoting peaceful change, and the only one which was ever invoked for the purpose,[1] was not Article 19 but Article 15, under which recommendations could be made without the concurrence of the parties concerned, and might, in the event of war, be supported by sanctions. But the only condition which could bring this article into operation was a " dispute likely to lead to a rupture ". The grievances of which the Covenant took cognisance were, broadly speaking, the grievances of states strong enough to create a danger of war. In 1932, when Finland brought before the Council a claim against Great Britain arising out of the commandeering of Finnish ships in the first world war, the British Government argued *inter alia* that there was no case to go to the Council, since the dispute was not in the least " likely to lead to a rupture ". In the same year, the British Government brought before the Council under Article 15 a dispute with Iran arising out of the affairs of the Anglo-Iranian Oil Company. The essential difference was that Great Britain was strong enough to create the danger of a rupture, and Finland was not.[2] When Article 19

If Czecho-Slovak territory had been transferred to Germany in September 1938 by a decision of the League Assembly or of an equity tribunal, enforced by mobilising the armies of the League or an international police force, the change would not for that reason have had any better title to the epithet " peaceful ". Armed force would have been used in precisely the same way.

[1] The Special Assembly, dealing with the Manchurian dispute under Article 15 of the Covenant, endorsed the recommendations of the Lytton Commission for substantial modifications of the *status quo* in Manchuria. It need hardly be added that Japanese military action was the force which prompted these recommendations, which proved, however, insufficient to satisfy Japan.

[2] In the Finnish question, M. Madariaga expressed the view that " it was extremely dangerous for the Council, the Assembly, and the League of Nations

was invoked for the first time by Bolivia in 1921, it was cogently argued that, since the conditions of which Bolivia complained had existed for a long period without endangering peace, there was no case for bringing them before the League. In other words, it would have been necessary, in order to set the procedure of peaceful change in motion, that Bolivia should be strong enough to threaten war against Chile. The doctrine of the Covenant thus confirmed the lesson of experience that peaceful change could not be effected on any important scale in international politics in the absence of a threat, or potential threat, of war.

We may sum up the conclusions so far reached. The judicial process is unfitted to solve the problem of peaceful change in national, and *a fortiori* in international, politics ; for, treating the parties to a dispute as equal, it fails to recognise the element of power which is a necessary factor in every demand for change. The legislative process, though recognising the role of power and well adapted to meet many demands for change in national politics, is inapplicable to international demands for change, since it presupposes the existence of a legislative authority whose decrees are binding on all members of the community without their specific assent. There remains the bargaining process, which is applied to some demands for change within the state and is alone applicable to demands for international change, since states (like trade unions or employers' federations) insist on the ultimate right to accept or reject any solution offered. But whereas under the legislative process change is enforced by the power of the state, change under the bargaining process can be enforced only by the power of the complainant. Power, used, threatened or silently held in reserve, is an essential factor in international change ; and change will, generally speaking, be effected only in the interests of those by whom, or on whose behalf, power can be invoked. " Yielding to threats of force " is a normal part of the process of peaceful change.

This is one side of the picture ; and since it is the side which is ignored in most modern writing about international

to establish the doctrine that irascible parties would be listened to, and calm parties would not, because in the latter case there would be no question of a rupture " (*League of Nations : Official Journal*, November 1934, p. 1458). The defect of the Finnish case was, however, not so much that Finland was calm as that she was weak.

politics, it has been deliberately emphasised here. Nor should we underrate the value of peaceful change even considered solely from this point of view. If relations between employers and workers are such that the former cannot resist, or the latter cannot sustain, a demand for an increase in wages and a reduction in hours, it is preferable (quite apart from any question of the justice or injustice of the demand) that it should be conceded or rejected as the result of peaceful negotiations rather than as the result of a long and embittered strike which half ruins both employers and workers. If the relations of power between the leading European countries in 1877 made it inevitable that Bulgaria should be deprived of much of the territory allocated to her by the Treaty of San Stefano, then it was preferable that this result should be brought about by discussions round a table in Berlin rather than by a war between Great Britain and Austria-Hungary on the one side and Russia on the other. If we consider peaceful change merely as a more or less mechanical device, replacing the alternative device of war, for readjusting the distribution of territory and of other desirable things to changes in the equilibrium of political forces, it performs a function whose utility it would be hypocritical to deny. Many changes made in national communities, whether by legislation or otherwise, and recognised as salutary, have no other basis than this.

The Role of Morality in Political Change

Nevertheless, it is clear that there is another aspect of peaceful change which occupies men's thoughts, and that it is no more possible to discuss peaceful change than to discuss any other kind of political procedure in terms of power alone. When a contested demand for change is made, the question which immediately exercises the minds of most people is whether it is just. It is true that our view of its justice is likely to be coloured, and may be wholly determined, by our own interest. It is true that, if our interest is not strongly engaged, we shall be tempted to discover reasons for regarding as just a solution which seems inevitable, or which could only be avoided by a great effort on our part. It is also true that, here as in every other political issue, power plays a part in determining our moral outlook, so that we shall be disposed, other things being equal, to regard a solution desired by the strong or the many as juster than a solution desired by the weak or the few. But when all these allowances have been made, the view taken of the morality of the transaction — a view not wholly determined by considerations of power — will influence the attitude of the mass of people affected by it. " If orderly government is to command general assent ", writes Mr. Bertrand Russell, " some way must be found of persuading a majority of mankind to agree upon some doctrine other than that of Thrasymachus " ; [1] and if an orderly procedure of peaceful change is ever to be established in international relations, some way must be found of basing its operation not on power alone, but on that uneasy compromise between power and morality which is the foundation of all political life. The establishment of a procedure of peaceful negotiation in disputes between capital and labour presupposes, not merely an acute perception on both sides of the strength and weakness of their respective positions at any given time, but also a certain measure of common feeling as to what is just and reasonable in their mutual relations, a spirit of give-and-take and even of potential self-sacrifice, so that a basis, however imperfect, exists for discussing demands on grounds of justice recognised by both. It is the embryonic character of this common feeling between nations, not the lack of a world legislature, and not the insistence of states on being judges in their own cause, which is the real obstacle in the way of an international procedure of peaceful change.

How far is this common feeling operative in relation to demands for international change ? Clearly in some degree. Two concrete cases of demands for change may be selected for analysis, one from the quasi-international, the other from the international sphere.

In the nineteenth century, the demand for home rule for Ireland found among a large number of people in Great Britain a support based not on considerations of power, but on common recognition as a canon of international morality of the right of " oppressed nationalities " to self-determination and on a certain readiness to sacrifice self-interest to it. The stock of common feeling between Great Britain and Ireland was considerably greater than that commonly existing between two foreign

[1] B. Russell, *Power*, p. 100.

countries. Nevertheless, the demand for change did not become effective until, owing to the diversion of British military strength elsewhere, force could be placed behind it. If the compromise eventually arrived at in 1921 could have been achieved in 1916, it would have been a true example of peaceful change achieved, like most international examples of peaceful change, under threat of war. But even in 1921, the settlement could not have been reached, and above all could not have been lasting, solely on a basis of power. The Anglo-Irish Treaty was a flagrant case of " yielding to threats of force " : it was concluded with the authors of a successful rebellion. But it had its necessary moral foundation in the acceptance of a common standard of what was just and reasonable in mutual relations between the two countries, and in the readiness of both (and particularly of the stronger) to make sacrifices in the interest of conciliation ; and this made a striking success of an agreement about which the gloomiest prognostications were current at the time of its conclusion.

The second example is the failure to achieve a peaceful settlement with Germany in the period between the two world wars. The mass of political opinion in Great Britain and Germany (and in most other countries) agreed for many years that a criterion of justice and injustice could properly be applied to the Versailles Treaty ; and there was a surprisingly considerable, though far from complete, consensus of opinion about the parts of it which were just and unjust respectively. Unfortunately, Germany was almost wholly deficient for fifteen years after 1918 in that power which is, as we have seen, a necessary motive force in political change ; and this deficiency prevented effect being given, except on a minor scale, to the widespread consensus of opinion that parts of the Versailles Treaty ought to be modified. By the time Germany regained her power, she had adopted a completely cynical attitude about the role of morality in international politics. Even though she continued to base her claims on grounds of justice, she expressed them more and more clearly in terms of naked force ; and this reacted on the opinion of the *status quo* countries, which became more and more inclined to forget earlier admissions of the injustices of the Versailles Treaty and to consider the issue as exclusively one of power.[1] The easy acquiescence of the

[1] This reaction was, of course, intensified by Nazi Germany's domestic policy.

status quo Powers in such actions as the denunciation of the military clauses, the reoccupation of the Rhineland or the annexation of Austria was due, not wholly to the fact that it was the line of least resistance, but in part also to a consensus of opinion that these changes were in themselves reasonable and just.[1] Yet they were greeted in each case by official censures and remonstrances which inevitably created the impression that the remonstrating Powers acquiesced merely because they were unable or unwilling to make the effort to resist. Successive removals of long recognised injustices of the Versailles Treaty became a cause not of reconciliation, but of further estrangement, between Germany and the Versailles Powers, and destroyed instead of increasing the limited stock of common feeling which had formerly existed.

It is beyond the scope of the present book to discuss the present or future foreign policy of Great Britain or of any other state. But the defence of the *status quo* is not a policy which can be lastingly successful. It will end in war as surely as rigid conservatism will end in revolution. " Resistance to aggression ", however necessary as a momentary device of national policy, is no solution ; for readiness to fight to prevent change is just as unmoral as readiness to fight to enforce it. To establish methods of peaceful change is therefore the fundamental problem of international morality and of international politics. We can discard as purely utopian and muddle-headed plans for a procedure of peaceful change dictated by a world legislature or a world court. We can describe as utopian in the right sense (i.e. performing the proper function of a utopia in proclaiming an ideal to be aimed at, though not wholly attainable) the desire to eliminate the element of power and to base the bargaining process of peaceful change on a common feeling of what is just and reasonable. But we shall also keep in mind the realist view of peaceful change as an adjustment to the changed relations of power ; and since the party which is able to bring most power to bear normally emerges successful from operations of peaceful change, we shall do our best to make ourselves as powerful as we can. In practice, we know

---

[1] In Great Britain, a perusal of the British press for March 7 and 8, 1936, will shew how widely the reoccupation of the Rhineland was not merely tolerated, but welcomed. Subsequently, the tone of the press became less favourable, being manifestly influenced by the more critical official attitude.

that peaceful change can only be achieved through a compromise between the utopian conception of a common feeling of right and the realist conception of a mechanical adjustment to a changed equilibrium of forces. That is why a successful foreign policy must oscillate between the apparently opposite poles of force and appeasement.

CONCLUSION

CHAPTER 14

THE PROSPECTS OF A NEW
INTERNATIONAL ORDER

The End of the Old Order

PERIODS of crisis have been common in history. The characteristic feature of the crisis of the twenty years between 1919 and 1939 was the abrupt descent from the visionary hopes of the first decade to the grim despair of the second, from a utopia which took little account of reality to a reality from which every element of utopia was rigorously excluded. The mirage of the nineteen-twenties was, as we now know, the belated reflexion of a century past beyond recall — the golden age of continuously expanding territories and markets, of a world policed by the self-assured and not too onerous British hegemony, of a coherent " Western " civilisation whose conflicts could be harmonised by a progressive extension of the area of common development and exploitation, of the easy assumptions that what was good for one was good for all and that what was economically right could not be morally wrong. The reality which had once given content to this utopia was already in decay before the nineteenth century had reached its end. The utopia of 1919 was hollow and without substance. It was without influence on the future because it no longer had any roots in the present.

The first and most obvious tragedy of this utopia was its ignominious collapse, and the despair which this collapse brought with it. " The European masses realised for the first time ", said a writer before the second world war, " that existence in this society is governed not by rational and sensible, but by blind, irrational and demonic forces." [1] It was no longer possible to rationalise international relations by pretending that what was good for Great Britain was also good for Yugoslavia and what was good for Germany was also good for Poland,

---

[1] P. Drucker, *The End of Economic Man*, p. 56.

so that international conflicts were merely the transient products of avoidable misunderstanding or curable ill-will. For more than a hundred years, the reality of conflict had been spirited out of sight by the political thinkers of Western civilisation. The men of the nineteen-thirties returned shocked and bewildered to the world of nature. The brutalities which, in the eighteenth and nineteenth centuries, were confined to dealings between civilised and uncivilised peoples were turned by civilised peoples against one another. The relation of totalitarianism to the crisis was clearly one not of cause, but of effect. Totalitarianism was not the disease, but one of the symptoms. Wherever the crisis raged, traces of this symptom could be found.

The second tragedy of the collapse of utopia, which proceeded from the first and further intensified it, was of a subtler kind. In the latter half of the nineteenth century, when the harmony of interests was already threatened by conflicts of increasing gravity, the rationality of the world was saved by a good stiff dose of Darwinism. The reality of conflict was admitted. But since conflict ended in the victory of the stronger, and the victory of the stronger was a condition of progress, honour was saved at the expense of the unfit. After 1919 only Fascists and Nazis clung openly to this outmoded device for rationalising and moralising international relations. But the Western countries resorted to an equally dubious and disastrous expedient. Smitten by the bankruptcy of the harmony of interests, and shocked by its Darwinian deviation, they attempted to build up a new international morality on the foundation, not of the right of the stronger, but of the right of those in possession. Like all utopias which are institutionalised, this utopia became the tool of vested interests and was perverted into a bulwark of the *status quo*. It is a moot point whether the politicians and publicists of the satisfied Powers, who attempted to identify international morality with security, law and order and other time-honoured slogans of privileged groups, do not bear their share of responsibility for the disaster as well as the politicians and publicists of the dissatisfied Powers, who brutally denied the validity of an international morality so constituted. Both these attempts to moralise international relations necessarily failed. We can accept neither the Darwinian doctrine, which identifies the good of the whole with the good of the fittest and contemplates without repugnance

Mussolini, "so similar inequality between the nations is calculated, if not peaceably adjusted, to lead to explosions of a much graver character."[1] The new harmony which was required was not (as the philosophers of *laissez-faire* assumed) a harmony between individuals, and not (as Marx assumed when he denied the possibility of its realisation) a harmony between classes, but a harmony between nations. To-day we need not make the mistake, which Marx made about the social class, of treating the nation as the ultimate group unit of human society. We need not pause to argue whether it is the best or the worst kind of unit to serve as the focus of political power. But we are bound to ask ourselves whether, and if so by what, it is likely to be superseded. Speculation on this subject falls naturally into two questions :

(*a*) Are the largest and most comprehensive units of political power in the world necessarily of a territorial character ?

(*b*) If so, will they continue to take approximately the form of the contemporary nation-state ?

The question whether the largest and most comprehensive power units must necessarily be territorial cannot receive a dogmatic answer applicable to all periods of history. At present, such units have a distinctively territorial form. It is easy to read past history as a gradual development leading up, with occasional relapses, to this consummation ; and political power is probably never, even in the most primitive societies, entirely divorced from the possession of territory. Yet in many periods of history, of which the mediaeval is the most recent, power has been based ostensibly — and in part, really — on grounds other than those of territorial sovereignty. It was acceptance of the principle *cuius regio eius religio* which substituted the unit based on domicile for the unit based on religious allegiance, and thereby laid the foundation of the modern nation-state. In no previous period of modern history have frontiers been so rigidly demarcated, or their character as barriers so ruthlessly enforced, as to-day ; and in no period, as we have already seen, has it been apparently so impossible to organise and maintain any international form of power. Modern technique, military and economic, seems to have indissolubly welded together power and territory. It is difficult for

[1] *The Times*, April 21, 1939.

contemporary man even to imagine a world in which political power would be organised on a basis not of territory, but of race, creed or class. Yet the enduring appeal of ideologies which transcend the limits of existing political units cannot be ignored. Few things are permanent in history ; and it would be rash to assume that the territorial unit of power is one of them. Its abandonment in favour of some other form of organised group power would, however, be so revolutionary that little that holds true of international politics in the present period would apply to the new dispensation. International relations would be supplanted by a new set of group relationships.

The question whether the territorial units of the future are likely to retain approximately their present form is one of more immediate practical importance. The problem of the optimum size of units — whether units of industrial or agricultural production or units of political and economic power — is one of the most puzzling and important of the present time ; and the near future may well see striking developments. In the field of political power, two contrary tendencies may be observed.

In one direction, there is a clearly marked trend towards integration and the formation of ever larger political and economic units. This trend set in in the latter part of the nineteenth century, and appears to have been closely connected with the growth of large-scale capitalism and industrialism, as well as with the improvement of means of communication and of the technical instruments of power. The first world war threw this development into conspicuous relief.

> Sovereignty, that is freedom to make decisions of wide historical importance [wrote Naumann in his famous book published in 1915], is now concentrated at a very few places on the globe. The day is still distant when there shall be " one fold, one shepherd ", but the days are past when shepherds without number, lesser or greater, drove their flocks unrestrained over the pastures of Europe. The spirit of large-scale industry and of super-national organisation has seized politics. . . . This is in conformity with centralised military technique.[1]

---

[1] F. Naumann, *Central Europe* (Engl. transl.), pp. 4-5.

Conclusion

The interlude of 1918, when nationalism momentarily resumed its disintegrating role, proved — at any rate in Europe — a dangerous fiasco. The multiplication of economic units added disastrously to the problems of the post-war period. Naumann with his *Mittel-Europa* proved a surer prophet than Woodrow Wilson with his principle of self-determination. The victors of 1918 " lost the peace " in Central Europe because they continued to pursue a principle of political and economic disintegration in an age which called for larger and larger units. The process of concentration still continued. The more autarky is regarded as the goal, the larger the units must become. The United States strengthened their hold over the American Continents. Great Britain created a " sterling *bloc* " and laid the foundations of a closed economic system. Germany reconstituted *Mittel-Europa* and pressed forward into the Balkans. Soviet Russia developed its vast territories into a compact unit of industrial and agricultural production. Japan attempted the creation of a new unit of " Eastern Asia " under Japanese domination. Such was the trend towards the concentration of political and economic power in the hands of six or seven highly organised units, round which lesser satellite units revolved without any appreciable independent motion of their own. On the other hand there is some evidence that, while technical, industrial and economic development within the last hundred years has dictated a progressive increase in the size of the effective political unit, there may be a size which cannot be exceeded without provoking a recrudescence of disintegrating tendencies. If any such law is at work, it is impossible to formulate it with any precision ; and prolonged investigation would be necessary to throw any light on the conditions which govern the size of political and economic units. The issue is, however, perhaps likely to be more decisive than any other for the course of world history in the next few generations.

One prediction may be made with some confidence. The concept of sovereignty is likely to become in the future even more blurred and indistinct than it is at present. The term was invented after the break-up of the mediaeval system to describe the independent character of the authority claimed and exercised by states which no longer recognised even the formal overlordship of the Empire. It was never more than a convenient label ; and when distinctions began to be made between political, legal and economic sovereignty or between internal and external sovereignty, it was clear that the label had ceased to perform its proper function as a distinguishing mark for a single category of phenomena. Discussion of such questions as whether the British Dominions are " sovereign " Powers, or in whom " sovereignty " of the mandated territories is vested, reveals the growing confusion. Such discussions are either legal arguments on the question what powers the authorities in those areas are constitutionally entitled to exercise (in which case the use of the term " sovereignty " gives little help), or else arguments of pure form on the question whether it is convenient to use the label " sovereignty " to describe situations which diverge to a greater or less extent from a common pattern. The concept of sovereignty becomes definitely misleading when, for instance, in a computation of the value of British colonial trade or British colonial investment, Egypt and Iraq are excluded on the ground that they are sovereign states. It is unlikely that the future units of power will take much account of formal sovereignty. There is no reason why each unit should not consist of groups of several formally sovereign states so long as the effective (but not necessarily the nominal) authority is exercised from a single centre. The effective group unit of the future will in all probability not be the unit formally recognised as such by international law. Any project of an international order which takes these formal units as its basis seems likely to prove unreal.

It may be well to add at this point that group units in some form will certainly survive as repositories of political power, whatever form these units may take. Nationalism was one of the forces by which the seemingly irreconcilable clash of interest between classes within the national community was reconciled. There is no corresponding force which can be invoked to reconcile the now seemingly irreconcilable clash of interest between nations. It is profitless to imagine a hypothetical world in which men no longer organise themselves in groups for purposes of conflict ; and the conflict cannot once more be transferred to a wider and more comprehensive field. As has often been observed, the international community cannot be organised against Mars. This is merely another aspect of the dilemma with which the collapse of the spacious conditions of nineteenth-century civilisation has confronted us. It seems no longer possible to create an apparent harmony of interests at the expense of somebody else. The conflict can no longer be spirited away.

Power in the New International Order

Power is a necessary ingredient of every political order. Historically, every approach in the past to a world society has been the product of the ascendancy of a single Power. In the nineteenth century the British fleet not only guaranteed immunity from major wars, but policed the high seas and offered equal security to all ; the London money market established a single currency standard for virtually the whole world ; British commerce secured — it is true, in an imperfect and attenuated form — a widespread acceptance of the principle of free trade ; and English became the *lingua franca* of four continents. These conditions, which were at once the product and the guarantee of British supremacy, created the illusion — and to some extent the reality — of a world society possessing interests and sympathies in common. The working hypothesis of an international order was created by a superior power. The hypothesis has been destroyed by the decline, relative or absolute, of that power. The British fleet is no longer strong enough to prevent war ; the London market can enforce a single currency standard only over a limited area ; free trade has wholly broken down ; and if the English language retains, and has increased, its ascendancy, this is due to the fact that it is shared by Great Britain with other important countries. By what power can the international order be restored ?

This question is likely to be answered by different nations in different ways. Most contemporary Englishmen are aware that the conditions which secured the overwhelming ascendancy of Great Britain in the nineteenth century no longer exist. But they sometimes console themselves with the dream that British supremacy, instead of passing altogether away, will be transmuted into the higher and more effective form of an ascendancy of the English-speaking peoples. The *pax Britannica* will be put into commission and become a *pax Anglo-Saxonica*, under which the British Dominions, standing half-way between the mother country and the United States, will be cunningly woven into a fabric of Anglo-American co-operation. This romantic idea goes back to the last years of the nineteenth century when Great Britain was already conscious of the growing burden of world supremacy, and when Cecil Rhodes had one of the first recorded visions of world empire based on an Anglo-American partnership. Oddly enough, it was an American Ambassador in London who, just before the war, gave the idea its most concrete expression. In 1913, Walter Hines Page proposed that President Wilson should visit London and conclude an Anglo-American alliance. " I think ", he added, " the world would take notice to whom it belongs and — be quiet." [1] The Washington Naval Treaty of 1922 was a more or less conscious bid by Great Britain for an equal partnership with the United States in the management of the world. The hope was reiterated again and again, with the reserves and the caution dictated by American susceptibilities, by British statesmen between the two world wars.

> I have always believed [said Lord Baldwin at the Albert Hall in May 1935] that the greatest security against war in any part of the world whatever, in Europe, in the East, anywhere, would be the close collaboration of the British Empire with the United States of America. The combined powers of the navies, the potential man-power, the immediate economic power of the combined blockade, and a refusal to trade or lend money would be a sanction that no power on earth however strong dare face. It may be a hundred years before that desirable end may be attained ; it may never come to pass. But sometimes we may have our dreams. I look forward to the future, and I see that union of forces for peace and justice in the world, and I cannot but think, even if men cannot advocate it openly yet, that some day and some time those who follow us may see it and know that the peace of the world is guaranteed by those who speak our tongue. [2]

The enormous growth of interest in Great Britain in everything relating to the United States shews what deep roots this ambition has struck in British hearts.

On the other side of the Atlantic, the picture necessarily looks rather different. Instead of an old firm, anxious to renew its strength by taking young blood into partnership,

---

[1] R. S. Baker, *Woodrow Wilson. Life and Letters*, v. p. 31.
[2] *The Times*, May 28, 1935.

we have here a young and untried nation, reliant on its own strength, but still uncertain how far that strength will carry it. The United States did not, until the turn of the century, stake out their claim for recognition as a Great Power. But it was not long before leading Americans were beginning to see visions of world supremacy.

> My dream [said Woodrow Wilson in a speech on Independence Day, 1914] is that as the years go by and the world knows more and more of America, it . . . will turn to America for those moral inspirations which lie at the basis of all freedom, . . . and that America will come into the full light of day when all shall know that she puts human rights above all other rights, and that her flag is the flag not only of America, but of humanity.[1]

The dream proved prophetic. In 1918 world leadership was offered, by almost unanimous consent, to the United States. The fact that it was then declined does not prove that it may not be grasped at some future time. If historical precedents count for anything, a *pax Americana* imposed on a divided and weakened Europe would be an easier contingency to realise than a *pax Anglo-Saxonica* based on an equal partnership of English-speaking peoples. But we are here in the realm of speculation, where the serious student cannot do more than canvass guesses and possibilities.

The necessary drawback about all conceptions of a world order depending on the ascendancy of a superior Power is that they ultimately involve recognition of the right of the strongest to assume world leadership. The *pax Romana* was the product of Roman imperialism, the *pax Britannica* of British imperialism. The " good neighbour " policy of the United States in Latin America is not the antithesis, but the continuation and consequence, of " Yankee imperialism " ; for it is only the strongest who can both maintain their supremacy and remain " good neighbours ". There is no theoretical reason to refuse to other nations the right to aspire to world leadership.

> Whoever really desires in his heart the victory of the pacifist conception of the world [writes Hitler in *Mein Kampf*] must devote himself by every means to the conquest of the

---

[1] R. S. Baker, *Woodrow Wilson and World Settlement*, i. p. 18.

world by the Germans. . . . The pacifist, humanitarian idea will perhaps be excellent when the man superior to all others shall first have so conquered and subjugated the world that he becomes its sole master.[1]

The policy of Japan, as the Chinese delegate remarked at an Assembly of the League of Nations, was to establish a *pax Japonica* in the Far East.[2] The Englishman or the American is entitled to resist such ambitions. But he cannot resist them on universal grounds which will appeal to the German or the Japanese. The conception of a *pax Germanica* or a *pax Japonica*, i.e. of a world order dominated by Germany or Japan, was *a priori* no more absurd and presumptuous than the conception of a *pax Britannica* would have seemed in the reign of Elizabeth or of a *pax Americana* in the days of Washington and Madison. The only reason why it would seem absurd for Nicaragua or Lithuania to aspire to world leadership is that, according to any reasonable prognostication, these countries will never be strong enough to have the slightest hope of attaining such an ambition. To attempt to ignore power as a decisive factor in every political situation is purely utopian. It is scarcely less utopian to imagine an international order built on a coalition of states, each striving to defend and assert its own interests. The new international order can be built only on a unit of power sufficiently coherent and sufficiently strong to maintain its ascendancy without being itself compelled to take sides in the rivalries of lesser units. Whatever moral issues may be involved, there is an issue of power which cannot be expressed in terms of morality.

Morality in the New International Order

If, however, it is utopian to ignore the element of power, it is an unreal kind of realism which ignores the element of morality in any world order. Just as within the state every government, though it needs power as a basis of its authority, also needs the moral basis of the consent of the governed, so an international order cannot be based on power alone, for the simple reason that mankind will in the long run always

[1] Hitler, *Mein Kampf*, p. 315.
[2] *League of Nations : Eighteenth Assembly*, p. 49.

revolt against naked power. Any international order pre-supposes a substantial measure of general consent. We shall, indeed, condemn ourselves to disappointment if we exaggerate the role which morality is likely to play. The fatal dualism of politics will always keep considerations of morality entangled with considerations of power. We shall never arrive at a political order in which the grievances of the weak and the few receive the same prompt attention as the grievances of the strong and the many. Power goes far to create the morality convenient to itself, and coercion is a fruitful source of consent. But when all these reserves have been made, it remains true that a new international order and a new international harmony can be built up only on the basis of an ascendancy which is generally accepted as tolerant and unoppressive or, at any rate, as preferable to any practicable alternative. To create these conditions is the moral task of the ascendant Power or Powers. The most effective moral argument which could be used in favour of a British or American, rather than a German or Japanese, hegemony of the world was that Great Britain and the United States, profiting by a long tradition and by some hard lessons in the past, have on the whole learned more successfully than Germany and Japan the capital importance of this task. Belief in the desirability of seeking the consent of the governed by methods other than those of coercion has in fact played a larger part in the British and American than in the German or Japanese administration of subject territories. Belief in the uses of conciliation even in dealing with those against whom it would have been easy to use force has in the past played a larger part in British and American than in German and Japanese foreign policy. That any moral superiority which this may betoken is mainly the product of long and secure enjoyment of superior power does not alter the fact, though this consideration may well affect the appeal of the argument to Germans and Japanese and expose British and Americans to the charge of self-righteousness when they invoke it.

It is, however, useless to discuss these problems of power and morality in a nineteenth-century setting, as if some fortunate turn of the wheel could restore the old conditions and allow a reconstitution of the international order on something like the old lines. The real international crisis of the modern world is the final and irrevocable breakdown of the conditions which made the nineteenth-century order possible. The old order cannot be restored, and a drastic change of outlook is unavoidable. Those who seek international conciliation may study with advantage the conditions which have made the process of conciliation between social classes in some degree successful. Essential conditions of that process were that the reality of the conflict should be frankly recognised, and not dismissed as an illusion in the minds of wicked agitators ; that the easy hypothesis of a natural harmony of interests, which a modicum of good-will and common sense would suffice to maintain, should be consigned to oblivion ; that what was morally desirable should not be identified with what was economically advantageous ; and that economic interests should, if necessary, be sacrificed in order to resolve the conflict by the mitigation of inequalities. None of these conditions has yet been realised in the international community. Responsible British and American statesmen still commonly speak as if there were a natural harmony of interests between the nations of the world which requires only good-will and common sense for its maintenance, and which is being wilfully disturbed by wicked dictators. British and American economists still commonly assume that what is economically good for Great Britain or the United States is economically good for other countries and therefore morally desirable. Few people are yet willing to recognise that the conflict between nations like the conflict between classes cannot be resolved without real sacrifices, involving in all probability a substantial reduction of consumption by privileged groups and in privileged countries. There may be other obstacles to the establishment of a new international order. But failure to recognise the fundamental character of the conflict, and the radical nature of the measures necessary to meet it, is certainly one of them.

Ultimately the best hope of progress towards international conciliation seems to lie along the path of economic reconstruction. Within the national community, necessity has carried us far towards the abandonment of economic advantage as the test of what is desirable. In nearly every country (and not least in the United States), large capital investments have been made in recent years, not for the economic purpose of earning profits, but for the social purpose of creating employment.

For some time the prejudice of orthodox economists against this policy was strong enough to restrict it to half measures. In Soviet Russia, such prejudice was non-existent from the outset. In the other totalitarian states, it rapidly disappeared. But elsewhere rearmament and war provided the first substantial cure for unemployment. The lesson will not be overlooked A repetition of the crisis of 1930–33 will not be tolerated anywhere, for the simple reason that workers have learned that unemployment can be cured by a gigantic programme of economically unremunerative expenditure on armaments ; and such expenditure would be equally effective from the standpoint of employment if it were devoted to some other economically unremunerative purpose such as the provision of free housing, free motor cars or free clothing. In the meanwhile we are moving rapidly everywhere towards the abolition or restriction of industrial profits. In the totalitarian countries this has now been virtually accomplished. In Great Britain, the assumption has long been made that to earn more than a limited rate of profit on the provision of essential public services is immoral. This assumption has now been extended to the armaments industry. Its extension to other industries is only a matter of time, and will be hastened by any crisis. The rearmament crisis of 1939, even if it had passed without war, would have produced everywhere changes in the social and industrial structure less revolutionary only than those produced by war itself. And the essence of this revolution is the abandonment of economic advantage as the test of policy. Employment has become more important than profit, social stability than increased consumption, equitable distribution than maximum production.

Internationally, this revolution complicates some problems and may help to solve others. So long as power wholly dominates international relations, the subordination of every other advantage to military necessity intensifies the crisis, and gives a foretaste of the totalitarian character of war itself. But once the issue of power is settled, and morality resumes its role, the situation is not without hope. Internationally as nationally, we cannot return to the pre-1939 world any more than we could return to the pre-war world in 1919. Frank acceptance of the subordination of economic advantage to social ends, and the recognition that what is economically good is not always morally good, must be extended from the national to the international sphere. The increasing elimination of the profit motive from the national economy should facilitate at any rate its partial elimination from foreign policy. After 1918, both the British and United States Governments granted to certain distressed countries " relief credits ", from which no economic return was ever seriously expected. Foreign loans for the purpose of stimulating production in export trades have been a familiar feature of post-war policy in many countries. Later extensions of this policy were dictated mainly by military considerations. But if the power crisis can be overcome, there can be no reason why it should not be extended for other purposes. The more we subsidise unproductive industries for political reasons, the more the provision of a rational employment supplants maximum profit as an aim of economic policy, the more we recognise the need of sacrificing economic advantage for social ends, the less difficult will it seem to realise that these social ends cannot be limited by a national frontier, and that British policy may have to take into account the welfare of Lille or Düsseldorf or Lodz as well as the welfare of Oldham or Jarrow. The broadening of our view of national policy should help to broaden our view of international policy ; and as has been said in an earlier chapter,[1] it is by no means certain that a direct appeal to the motive of sacrifice would always fail.

This, too, is a utopia. But it stands more directly in the line of recent advance than visions of a world federation or blue-prints of a more perfect League of Nations. Those elegant superstructures must wait until some progress has been made in digging the foundations.

[1] See p. 169.

INDEX

Index

Index

Index

THE END

harper ☿ torchbooks

HUMANITIES AND SOCIAL SCIENCES

American Studies: General

THOMAS C. COCHRAN: The Inner Revolution: *Essays on the Social Sciences in History*   TB/1140

EDWARD S. CORWIN: American Constitutional History. *Essays edited by Alpheus T. Mason and Gerald Garvey*   TB/1136

A. HUNTER DUPREE: Science in the Federal Government: *A History of Policies and Activities to 1940*   TB/573

OSCAR HANDLIN, Ed.: This Was America: *As Recorded by European Travelers in the Eighteenth, Nineteenth and Twentieth Centuries. Illus.*   TB/1119

MARCUS LEE HANSEN: The Atlantic Migration: 1607-1860. *Edited by Arthur M. Schlesinger; Introduction by Oscar Handlin*   TB/1052

MARCUS LEE HANSEN: The Immigrant in American History. *Edited with a Foreword by Arthur M. Schlesinger*   TB/1120

JOHN HIGHAM, Ed.: The Reconstruction of American History   TB/1068

ROBERT H. JACKSON: The Supreme Court in the American System of Government   TB/1106

JOHN F. KENNEDY: A Nation of Immigrants. *Illus. Revised and Enlarged. Introduction by Robert F. Kennedy*   TB/1118

RALPH BARTON PERRY: Puritanism and Democracy   TB/1138

ARNOLD ROSE: The Negro in America: *The Condensed Version of Gunnar Myrdal's An American Dilemma*   TB/3048

MAURICE R. STEIN: The Eclipse of Community: *An Interpretation of American Studies*   TB/1128

W. LLOYD WARNER and Associates: Democracy in Jonesville: *A Study in Quality and Inequality* || TB/1129

W. LLOYD WARNER: Social Class in America: *The Evaluation of Status*   TB/1013

American Studies: Colonial

BERNARD BAILYN, Ed.: The Apologia of Robert Keayne: *Self-Portrait of a Puritan Merchant*   TB/1201

BERNARD BAILYN: The New England Merchants in the Seventeenth Century   TB/1149

JOSEPH CHARLES: The Origins of the American Party System   TB/1049

LAWRENCE HENRY GIPSON: The Coming of the Revolution: 1763-1775. † *Illus.*   TB/3007

LEONARD W. LEVY: Freedom of Speech and Press in Early American History: *Legacy of Suppression*   TB/1109

PERRY MILLER: Errand Into the Wilderness   TB/1139

PERRY MILLER & T. H. JOHNSON, Eds.: The Puritans: *A Sourcebook of Their Writings*
Vol. I   TB/1093; Vol. II   TB/1094

KENNETH B. MURDOCK: Literature and Theology in Colonial New England   TB/99

WALLACE NOTESTEIN: The English People on the Eve of Colonization: 1603-1630. † *Illus.*   TB/3006

LOUIS B. WRIGHT: The Cultural Life of the American Colonies: 1607-1763. † *Illus.*   TB/3005

American Studies: From the Revolution to the Civil War

JOHN R. ALDEN: The American Revolution: 1775-1783. † *Illus.*   TB/3011

RAY A. BILLINGTON: The Far Western Frontier: 1830-1860. † *Illus.*   TB/3012

GEORGE DANGERFIELD: The Awakening of American Nationalism: 1815-1828. † *Illus.*   TB/3061

CLEMENT EATON: The Freedom-of-Thought Struggle in the Old South. *Revised and Enlarged. Illus.*   TB/1150

CLEMENT EATON: The Growth of Southern Civilization: 1790-1860. † *Illus.*   TB/3040

LOUIS FILLER: The Crusade Against Slavery: 1830-1860. † *Illus.*   TB/3029

DIXON RYAN FOX: The Decline of Aristocracy in the Politics of New York: 1801-1840. ‡ *Edited by Robert V. Remini*   TB/3064

FELIX GILBERT: The Beginnings of American Foreign Policy: *To the Farewell Address*   TB/1200

FRANCIS J. GRUND: Aristocracy in America: *Social Class in the Formative Years of the New Nation*   TB/1001

ALEXANDER HAMILTON: The Reports of Alexander Hamilton. ‡ *Edited by Jacob E. Cooke*   TB/3060

THOMAS JEFFERSON: Notes on the State of Virginia. ‡ *Edited by Thomas P. Abernethy*   TB/3052

BERNARD MAYO: Myths and Men: *Patrick Henry, George Washington, Thomas Jefferson*   TB/1108

JOHN C. MILLER: Alexander Hamilton and the Growth of the New Nation   TB/3057

RICHARD B. MORRIS, Ed.: The Era of the American Revolution   TB/1180

R. B. NYE: The Cultural Life of the New Nation: 1776-1801. † *Illus.*   TB/3026

† The New American Nation Series, edited by Henry Steele Commager and Richard B. Morris.

‡ American Perspectives series, edited by Bernard Wishy and William E. Leuchtenburg.

* The Rise of Modern Europe series, edited by William L. Langer.

|| Researches in the Social, Cultural, and Behavioral Sciences, edited by Benjamin Nelson.

§ The Library of Religion and Culture, edited by Benjamin Nelson.

υ Not for sale in Canada.

Σ Harper Modern Science Series, edited by James R. Newman.

3

WALLACE K. FERGUSON et al.: Facets of the Renaissance
TB/1098
WALLACE K. FERGUSON et al.: The Renaissance: Six Essays. Illus. TB/1084
JOHN NEVILLE FIGGIS: The Divine Right of Kings. Introduction by G. R. Elton TB/1191
JOHN NEVILLE FIGGIS: Political Thought from Gerson to Grotius: 1414-1625: Seven Studies. Introduction by Garrett Mattingly TB/1032
MYRON P. GILMORE: The World of Humanism, 1453-1517.* Illus. TB/3003
FRANCESCO GUICCIARDINI: Maxims and Reflections of a Renaissance Statesman (Ricordi). Trans. by Mario Domandi. Intro. by Nicolai Rubinstein TB/1160
J. H. HEXTER: More's Utopia: The Biography of an Idea New Epilogue by the Author TB/1195
JOHAN HUIZINGA: Erasmus and the Age of Reformation. Illus. TB/19
ULRICH VON HUTTEN et al.: On the Eve of the Reformation: "Letters of Obscure Men." Introduction by Hajo Holborn TB/1124
PAUL O. KRISTELLER: Renaissance Thought: The Classic, Scholastic, and Humanist Strains TB/1048
PAUL O. KRISTELLER: Renaissance Thought II: Papers on Humanism and the Arts TB/1163
NICCOLO MACHIAVELLI: History of Florence and of the Affairs of Italy: from the earliest times to the death of Lorenzo the Magnificent. Introduction by Felix Gilbert TB/1027
ALFRED VON MARTIN: Sociology of the Renaissance. Introduction by Wallace K. Ferguson TB/1099
GARRETT MATTINGLY et al.: Renaissance Profiles. Edited by J. H. Plumb TB/1162
MILLARD MEISS: Painting in Florence and Siena after the Black Death: The Arts, Religion and Society in the Mid-Fourteenth Century. 169 illus. TB/1148
J. E. NEALE: The Age of Catherine de Medici ⁰ TB/1085
ERWIN PANOFSKY: Studies in Iconology: Humanistic Themes in the Art of the Renaissance. 180 illustrations TB/1077
J. H. PARRY: The Establishment of the European Hegemony: 1415-1715: Trade and Exploration in the Age of the Renaissance TB/1045
J. H. PLUMB: The Italian Renaissance: A Concise Survey of Its History and Culture TB/1161
CECIL ROTH: The Jews in the Renaissance. Illus. TB/834
GORDON RUPP: Luther's Progress to the Diet of Worms ⁰ TB/120
FERDINAND SCHEVILL: The Medici. Illus. TB/1010
FERDINAND SCHEVILL: Medieval and Renaissance Florence. Illus. Volume I: Medieval Florence TB/1090 Volume II: The Coming of Humanism and the Age of the Medici TB/1091
G. M. TREVELYAN: England in the Age of Wycliffe, 1368-1520 ⁰ TB/1112
VESPASIANO: Renaissance Princes, Popes, and Prelates: The Vespasiano Memoirs: Lives of Illustrious Men of the XVth Century. Intro. by Myron P. Gilmore TB/1111

History: Modern European

FREDERICK B. ARTZ: Reaction and Revolution, 1815-1832. * Illus. TB/3034
MAX BELOFF: The Age of Absolutism, 1660-1815 TB/1062
ROBERT C. BINKLEY: Realism and Nationalism, 1852-1871. * Illus. TB/3038
ASA BRIGGS: The Making of Modern England, 1784-1867: The Age of Improvement ⁰ TB/1203
CRANE BRINTON: A Decade of Revolution, 1789-1799. * Illus. TB/3018

J. BRONOWSKI & BRUCE MAZLISH: The Western Intellectual Tradition: From Leonardo to Hegel TB/3001
GEOFFREY BRUUN: Europe and the French Imperium, 1799-1814. * Illus. TB/3033
ALAN BULLOCK: Hitler, A Study in Tyranny. ⁰ Illus. TB/1123
E. H. CARR: The Twenty Years' Crisis, 1919-1939: An Introduction to the Study of International Relations ⁰ TB/1122
GORDON A. CRAIG: From Bismarck to Adenauer: Aspects of German Statecraft. Revised Edition TB/1171
WALTER L. DORN: Competition for Empire, 1740-1763. * Illus. TB/3032
CARL J. FRIEDRICH: The Age of the Baroque, 1610-1660. * Illus. TB/3004
RENÉ FUELOEP-MILLER: The Mind and Face of Bolshevism: An Examination of Cultural Life in Soviet Russia. New Epilogue by the Author TB/1188
M. DOROTHY GEORGE: London Life in the Eighteenth Century TB/1182
LEO GERSHOY: From Despotism to Revolution, 1763-1789. * Illus. TB/3017
C. C. GILLISPIE: Genesis and Geology: The Decades before Darwin § TB/51
ALBERT GOODWIN: The French Revolution TB/1064
ALBERT GUERARD: France in the Classical Age: The Life and Death of an Ideal TB/1183
CARLTON J. H. HAYES: A Generation of Materialism, 1871-1900. * Illus. TB/3039
J. H. HEXTER: Reappraisals in History: New Views on History & Society in Early Modern Europe TB/1100
A. R. HUMPHREYS: The Augustan World: Society, Thought, and Letters in 18th Century England ⁰ TB/1105
ALDOUS HUXLEY: The Devils of Loudun: A Study in the Psychology of Power Politics and Mystical Religion in the France of Cardinal Richelieu § ⁰ TB/60
DAN N. JACOBS, Ed.: The New Communist Manifesto & Related Documents. Third edition, revised TB/1078
HANS KOHN: The Mind of Germany: The Education of a Nation TB/1204
HANS KOHN, Ed.: The Mind of Modern Russia: Historical and Political Thought of Russia's Great Age TB/1065
KINGSLEY MARTIN: French Liberal Thought in the Eighteenth Century: A Study of Political Ideas from Bayle to Condorcet TB/1114
SIR LEWIS NAMIER: Personalities and Powers: Selected Essays TB/1186
SIR LEWIS NAMIER: Vanished Supremacies: Essays on European History, 1812-1918 ⁰ TB/1088
JOHN U. NEF: Western Civilization Since the Renaissance: Peace, War, Industry, and the Arts TB/1113
FREDERICK L. NUSSBAUM: The Triumph of Science and Reason, 1660-1685. * Illus. TB/3009
JOHN PLAMENATZ: German Marxism and Russian Communism. ⁰ New Preface by the Author TB/1189
RAYMOND W. POSTGATE, Ed.: Revolution from 1789 to 1906: Selected Documents TB/1063
PENFIELD ROBERTS: The Quest for Security, 1715-1740. * Illus. TB/3016
PRISCILLA ROBERTSON: Revolutions of 1848: A Social History TB/1025
ALBERT SOREL: Europe Under the Old Regime. Translated by Francis H. Herrick TB/1121
N. N. SUKHANOV: The Russian Revolution, 1917: Eyewitness Account. Edited by Joel Carmichael Vol. I TB/1066; Vol. II TB/1067
A. J. P. TAYLOR: The Habsburg Monarch, 1809-1918: A History of the Austrian Empire and Austria-Hungary ⁰ TB/1187

4

JOHN B. WOLF: The Emergence of the Great Powers, 1685-1715. * Illus.　TB/3010

JOHN B. WOLF: France: 1814-1919: *The Rise of a Liberal-Democratic Society*　TB/3019

Intellectual History & History of Ideas

HERSCHEL BAKER: The Image of Man: *A Study of the Idea of Human Dignity in Classical Antiquity, the Middle Ages, and the Renaissance*　TB/1047

R. R. BOLGAR: The Classical Heritage and Its Beneficiaries: *From the Carolingian Age to the End of the Renaissance*　TB/1125

RANDOLPH S. BOURNE: War and the Intellectuals: *Collected Essays, 1915-1919.* ‡ *Edited by Carl Resek*　TB/3043

J. BRONOWSKI & BRUCE MAZLISH: The Western Intellectual Tradition: *From Leonardo to Hegel*　TB/3001

ERNST CASSIRER: The Individual and the Cosmos in Renaissance Philosophy. *Translated with an Introduction by Mario Domandi*　TB/1097

NORMAN COHN: The Pursuit of the Millennium: *Revolutionary Messianism in medieval and Reformation Europe*　TB/1037

G. RACHEL LEVY: Religious Conceptions of the Stone Age and Their Influence upon European Thought. *Illus. Introduction by Henri Frankfort*　TB/106

ARTHUR O. LOVEJOY: The Great Chain of Being: *A Study of the History of an Idea*　TB/1009

PERRY MILLER & T. H. JOHNSON, Editors: The Puritans: *A Sourcebook of Their Writings*
Vol. I TB/1093; Vol. II TB/1094

MILTON C. NAHM: Genius and Creativity: *An Essay in the History of Ideas*　TB/1196

ROBERT PAYNE: Hubris: *A Study of Pride. Foreword by Sir Herbert Read*　TB/1031

RALPH BARTON PERRY: The Thought and Character of William James: *Briefer Version*　TB/1156

BRUNO SNELL: The Discovery of the Mind: *The Greek Origins of European Thought*　TB/1018

PAGET TOYNBEE: Dante Alighieri: *His Life and Works. Edited with intro. by Charles S. Singleton*　TB/1206

ERNEST LEE TUVESON: Millennium and Utopia: *A Study in the Background of the Idea of Progress.* | *New Preface by the Author*　TB/1134

PAUL VALÉRY: The Outlook for Intelligence　TB/2016

PHILIP P. WIENER: Evolution and the Founders of Pragmatism. *Foreword by John Dewey*　TB/1212

Literature, Poetry, The Novel & Criticism

JAMES BAIRD: Ishmael: *The Art of Melville in the Contexts of International Primitivism*　TB/1023

JACQUES BARZUN: The House of Intellect　TB/1051

W. J. BATE: From Classic to Romantic: *Premises of Taste in Eighteenth Century England*　TB/1036

RACHEL BESPALOFF: On the Iliad　TB/2006

R. P. BLACKMUR et al.: Lectures in Criticism. *Introduction by Huntington Cairns*　TB/2003

ABRAHAM CAHAN: The Rise of David Levinsky: *a documentary novel of social mobility in early twentieth century America. Intro. by John Higham*　TB/1028

ERNST R. CURTIUS: European Literature and the Latin Middle Ages　TB/2015

GEORGE ELIOT: Daniel Deronda: *a novel. Introduction by F. R. Leavis*　TB/1039

ETIENNE GILSON: Dante and Philosophy　TB/1089

ALFRED HARBAGE: As They Liked It: *A Study of Shakespeare's Moral Artistry*　TB/1035

STANLEY R. HOPPER, Ed.: Spiritual Problems in Contemporary Literature §　TB/21

A. R. HUMPHREYS: The Augustan World: *Society, Thought and Letters in 18th Century England* °　TB/1105

ALDOUS HUXLEY: Antic Hay & The Giaconda Smile. ° *Introduction by Martin Green*　TB/3503

ALDOUS HUXLEY: Brave New World & Brave New World Revisited. ° *Introduction by Martin Green*　TB/3501

HENRY JAMES: Roderick Hudson: *a novel. Introduction by Leon Edel*　TB/1016

HENRY JAMES: The Tragic Muse: *a novel. Introduction by Leon Edel*　TB/1017

ARNOLD KETTLE: An Introduction to the English Novel. Volume I: *Defoe to George Eliot*　TB/1011
Volume II: *Henry James to the Present*　TB/1012

ROGER SHERMAN LOOMIS: The Development of Arthurian Romance　TB/1167

JOHN STUART MILL: On Bentham and Coleridge. *Introduction by F. R. Leavis*　TB/1070

KENNETH B. MURDOCK: Literature and Theology in Colonial New England　TB/99

SAMUEL PEPYS: The Diary of Samuel Pepys. ° *Edited by O. F. Morshead. Illus. by Ernest Shepard*　TB/1007

ST.-JOHN PERSE: Seamarks　TB/2002

GEORGE SANTAYANA: Interpretations of Poetry and Religion §　TB/9

C. P. SNOW: Time of Hope: *a novel*　TB/1040

HEINRICH STRAUMANN: American Literature in the Twentieth Century. *Third Edition, Revised*　TB/1168

PAGET TOYNBEE: Dante Alighieri: *His Life and Works. Edited with intro. by Charles S. Singleton*　TB/1206

DOROTHY VAN GHENT: The English Novel: *Form and Function*　TB/1050

E. B. WHITE: One Man's Meat. *Introduction by Walter Blair*　TB/3505

MORTON DAUWEN ZABEL, Editor: Literary Opinion in America　Vol. I TB/3013; Vol. II TB/3014

Myth, Symbol & Folklore

JOSEPH CAMPBELL, Editor: Pagan and Christian Mysteries Illus.　TB/2013

MIRCEA ELIADE: Cosmos and History: *The Myth of the Eternal Return* §　TB/2050

C. G. JUNG & C. KERÉNYI: Essays on a Science of Mythology: *The Myths of the Divine Child and the Divine Maiden*　TB/2014

DORA & ERWIN PANOFSKY: Pandora's Box: *The Changing Aspects of a Mythical Symbol. Revised Edition. Illus.*　TB/2021

ERWIN PANOFSKY: Studies in Iconology: *Humanistic Themes in the Art of the Renaissance. 180 illustrations*　TB/1077

JEAN SEZNEC: The Survival of the Pagan Gods: *The Mythological Tradition and its Place in Renaissance Humanism and Art. 108 illustrations*　TB/2004

HELLMUT WILHELM: Change: *Eight Lectures on the I Ching*　TB/2019

HEINRICH ZIMMER: Myths and Symbols in Indian Art and Civilization. *70 illustrations*　TB/2005

Philosophy

G. E. M. ANSCOMBE: An Introduction to Wittgenstein's Tractatus. *Second edition, Revised.* °　TB/1210

HENRI BERGSON: Time and Free Will: *An Essay on the Immediate Data of Consciousness* °　TB/1021

H. J. BLACKHAM: Six Existentialist Thinkers: *Kierke-gaard, Nietzsche, Jaspers, Marcel, Heidegger, Sartre* °
TB/1002

CRANE BRINTON: Nietzsche. *New Preface, Bibliography and Epilogue by the Author* TB/1197

ERNST CASSIRER: The Individual and the Cosmos in Renaissance Philosophy. *Translated with an Introduction by Mario Domandi* TB/1097

ERNST CASSIRER: Rousseau, Kant and Goethe. *Introduction by Peter Gay* TB/1092

FREDERICK COPLESTON: Medieval Philosophy ° TB/376

F. M. CORNFORD: Principium Sapientiae: *A Study of the Origins of Greek Philosophical Thought. Edited by W. K. C. Guthrie* TB/1213

F. M. CORNFORD: From Religion to Philosophy: *A Study in the Origins of Western Speculation* § TB/20

WILFRID DESAN: The Tragic Finale: *An Essay on the Philosophy of Jean-Paul Sartre* TB/1030

PAUL FRIEDLÄNDER: Plato: *An Introduction* TB/2017

ÉTIENNE GILSON: Dante and Philosophy TB/1089

WILLIAM CHASE GREENE: Moira: *Fate, Good, and Evil in Greek Thought* TB/1104

W. K. C. GUTHRIE: The Greek Philosophers: *From Thales to Aristotle* ° TB/1008

F. H. HEINEMANN: Existentialism and the Modern Predicament TB/28

ISAAC HUSIK: A History of Medieval Jewish Philosophy
TB/803

EDMUND HUSSERL: Phenomenology and the Crisis of Philosophy. *Translated with an Introduction by Quentin Lauer* TB/1170

IMMANUEL KANT: The Doctrine of Virtue, *being Part II of The Metaphysic of Morals. Trans. with Notes & Intro. by Mary J. Gregor. Foreword by H. J. Paton*
TB/110

IMMANUEL KANT: Groundwork of the Metaphysic of Morals. *Trans. & analyzed by H. J. Paton* TB/1159

IMMANUEL KANT: Lectures on Ethics. § *Introduction by Lewis W. Beck* TB/105

QUENTIN LAUER: Phenomenology: *Its Genesis and Prospect* TB/1169

GABRIEL MARCEL: Being and Having: *An Existential Diary. Intro. by James Collins* TB/310

GEORGE A. MORGAN: What Nietzsche Means TB/1198

PHILO SAADYA GAON, & JEHUDA HALEVI: Three Jewish Philosophers. *Ed. by Hans Lewy, Alexander Altmann, & Isaak Heinemann* TB/813

MICHAEL POLANYI: Personal Knowledge: *Towards a Post-Critical Philosophy* TB/1158

WILLARD VAN ORMAN QUINE: Elementary Logic: *Revised Edition* TB/577

WILLARD VAN ORMAN QUINE: From a Logical Point of View: *Logico-Philosophical Essays* TB/566

BERTRAND RUSSELL et al.: The Philosophy of Bertrand Russell. *Edited by Paul Arthur Schilpp*
Vol. I TB/1095; Vol. II TB/1096

L. S. STEBBING: A Modern Introduction to Logic TC/538

ALFRED NORTH WHITEHEAD: Process and Reality: *An Essay in Cosmology* TB/1033

PHILIP P. WIENER: Evolution and the Founders of Pragmatism. *Foreword by John Dewey* TB/1212

WILHELM WINDELBAND: A History of Philosophy
Vol. I: *Greek, Roman, Medieval* TB/38
Vol. II: *Renaissance, Enlightenment, Modern* TB/39

LUDWIG WITTGENSTEIN: The Blue and Brown Books °
TB/1211

Political Science & Government

JEREMY BENTHAM: The Handbook of Political Fallacies: *Introduction by Crane Brinton* TB/1069

KENNETH E. BOULDING: Conflict and Defense: *A General Theory* TB/3024

CRANE BRINTON: English Political Thought in the Nineteenth Century TB/1071

EDWARD S. CORWIN: American Constitutional History: *Essays edited by Alpheus T. Mason and Gerald Garvey* TB/1136

ROBERT DAHL & CHARLES E. LINDBLOM: Politics, Economics, and Welfare: *Planning and Politico-Economic Systems Resolved into Basic Social Processes* TB/3037

JOHN NEVILLE FIGGIS: The Divine Right of Kings. *Introduction by G. R. Elton* TB/1191

JOHN NEVILLE FIGGIS: Political Thought from Gerson to Grotius: 1414-1625: *Seven Studies. Introduction by Garrett Mattingly* TB/1032

F. L. GANSHOF: Feudalism TB/1058

G. P. GOOCH: English Democratic Ideas in Seventeenth Century TB/1006

J. H. HEXTER: More's Utopia: *The Biography of an Idea. New Epilogue by the Author* TB/1195

ROBERT H. JACKSON: The Supreme Court in the American System of Government TB/1106

DAN N. JACOBS, Ed.: The New Communist Manifesto & *Related Documents. Third edition, Revised* TB/1078

DAN N. JACOBS & HANS BAERWALD, Eds.: Chinese Communism: *Selected Documents* TB/3031

ROBERT GREEN MCCLOSKEY: American Conservatism in the Age of Enterprise, 1865-1910 TB/1137

KINGSLEY MARTIN: French Liberal Thought in the Eighteenth Century: *Political Ideas from Bayle to Condorcet* TB/1114

JOHN STUART MILL: On Bentham and Coleridge. *Introduction by F. R. Leavis* TB/1070

JOHN B. MORRALL: Political Thought in Medieval Times
TB/1076

JOHN PLAMENATZ: German Marxism and Russian Communism. ° *New Preface by the Author* TB/1189

SIR KARL POPPER: The Open Society and Its Enemies
Vol. I: *The Spell of Plato* TB/1101
Vol. II: *The High Tide of Prophecy: Hegel, Marx, and the Aftermath* TB/1102

HENRI DE SAINT-SIMON: Social Organization, The Science of Man, and Other Writings. *Edited and Translated by Felix Markham* TB/1152

JOSEPH A. SCHUMPETER: Capitalism, Socialism and Democracy TB/3008

CHARLES H. SHINN: Mining Camps: *A Study in American Frontier Government. ‡ Edited by Rodman W. Paul*
TB/3062

Psychology

ALFRED ADLER: The Individual Psychology of Alfred Adler. *Edited by Heinz L. and Rowena R. Ansbacher*
TB/1154

ALFRED ADLER: Problems of Neurosis. *Introduction by Heinz L. Ansbacher* TB/1145

ANTON T. BOISEN: The Exploration of the Inner World: *A Study of Mental Disorder and Religious Experience*
TB/87

HERBERT FINGARETTE: The Self in Transformation: *Psychoanalysis, Philosophy and the Life of the Spirit.* ‖
TB/1177

SIGMUND FREUD: On Creativity and the Unconscious: *Papers on the Psychology of Art, Literature, Love, Religion.* § *Intro. by Benjamin Nelson* TB/45

Christianity: General

Christianity: Origins & Early Development

Christianity: The Middle Ages and The Reformation

Christianity: The Protestant Tradition